WHEELED AFVS IN US SERVICE

One of the longest-running debates in the history of armoured warfare, indeed one that dates back to the introduction of the tank in 1916, concerns the relative merits of tracked armoured fighting vehicles (AFVs) over their wheeled counterparts. The tank, of course, was designed at first to meet a very specific operational need: to break the stalemate on the Western Front by introducing a vehicle that could cross No-Man's Land and protect the crew from the guns of the enemy while doing so. By the end of the Great War, the concept of the tank with a 360-degree rotating turret had been introduced and in the following decade the pioneers of armoured warfare evolved the trinity – protection, mobility and firepower – that determines the design of AFVs to the current day. The counterpart to the tank was in some ways the armoured car: a vehicle that sacrificed the mobility of caterpillar tracks but gained the speed of a wheeled vehicle, while maintaining some degree of the firepower and protection of the tank. The first armoured cars were, of course, principally used for reconnaissance and that role continued throughout the twentieth and into the twenty-first century as one of the primary roles of the wheeled AFV. By the 1930s a third type of AFV was actively being incorporated into the armoured warfare doctrines of several nations. This reflected the need for the infantry to keep pace with the tanks and the armoured cars in a vehicle that shared to some degree the mobility and protection, if not the firepower, of the rest of the developing AFV fleet. In World War II the most successful of these designs were, of course, the half-tracks – the German Sd.Kfz. 251 and the American M2/M3 – but they were by no means ubiquitous and most infantrymen walked into battle or were transported to the frontline in unarmoured lorries.

In the US Army the wheeled armoured car played an important role as reconnaissance vehicles throughout the war. These vehicles had been developed in the 1930s and when war began in 1941 the M3A1 Scout Car equipped the cavalry units of the US Army. More than 20,000 were built to 1944 and it also served in the British, Commonwealth armies and, through the Lend Lease programme, with the Red Army. In 1943 the M8 armoured car began production and began to equip the cavalry reconnaissance units, in which it played an important role in the European campaign as well as in the Philippines and on Okinawa. Over 12,000 were built between 1943 and 1945. The M8 and the M20 (a turretless version armed with a .50cal machine gun) continued to serve with US cavalry regiments into the Korean War.

Towards the end of World War II the US Army opted to pursue tracked armoured vehicles to transport infantry and fulfil other battlefield roles. This was in contrast to the Soviet Union who continued to develop wheeled armoured vehicles in both the armoured personnel carrier (APC) and reconnaissance roles. In 1952 production began of the first fully tracked American APC, the M75. This could carry an infantry squad into battle in a fully enclosed armoured and tracked vehicle. The M75 saw action in the closing stages of the Korean War and was superseded in 1953 by the much cheaper M59, of which some 6,300 were produced between 1953 and 1964. The M59 also had a mortar carrier version, M84, and proved to be a successful design. In the late 1950s the US Army began the development of an air-portable tracked armoured vehicle that would fulfil a number of battlefield roles, including APC and armoured reconnaissance.

This led, in 1959, to the production of the M113. The M113 was principally an APC, carrying eleven infantrymen into battle. It served in this role throughout the Vietnam War into the 1980s and continues to serve with the US Army in various roles to this day. The M113 was an incredibly versatile vehicle and was adapted to the armoured cavalry role, as a command vehicle, a mortar carrier, an anti-aircraft vehicle and in other specialist roles. Yet even as the M113 was entering service it was becoming clear that a new AFV was needed to survive the

An M3A1 Scout Car during training in the US. Between 1939 and 1944 some 20,000 of these vehicles were produced and they played an important role during World War II as a cavalry reconnaissance vehicle and a troop carrier with the Allied armies and with the Soviet Union. (US Department of Energy)

An M2 Bradley IFV is driven from one of the warehouses for Pre-Positioned Equipment in Germany during the annual Reforger exercise in 1984. (US Army photo by Spc. Vincent Kitts)

kind of high-intensity, Nuclear, Biological and Chemical (NBC) environment of any potential conflict with the Soviet Union. The US military recognised the need for a platform which would not only transport the infantry into battle but also allow them to engage the enemy in a fast-moving battle alongside the main battle tanks. Alongside the development of the MBT-70 between the Americans and Germans, there was also a programme to develop a Mechanized Infantry Combat Vehicle (MICV-70). Whereas the former led, in a convoluted way, to the Leopard 2 and M1 Abrams, the MICV project was abandoned in 1968 as the vehicle could not be airlifted. The first public display of the Soviet BMP-1 Infantry Fighting Vehicle (IFV) a year earlier had, however, underscored the need for a replacement for the M113. An IFV based on the M113 was rejected as being too slow, while the newly developed German Marder IFV was judged to be too heavy and expensive.

In 1972 the Ford Motor Company won a contract to develop a prototype IFV, the XM723. The vehicle had aluminium armour proof against projectiles up to 14.5mm in calibre and could carry a fully equipped infantry squad. It was originally armed with a turret-mounted 20mm cannon, but this was soon replaced with a two-man turret armed with a 25mm Bushmaster cannon and TOW missiles providing the all-important anti-armour capability. In 1977 it was re-named the XM2, with the cavalry reconnaissance version (which lacked the hull firing ports for the infantry squad and carried additional TOW missiles) named the XM3. The development of the XM2/3 was not straightforward and it faced political as well as military questioning over cost, size and its ability to survive on the NBC battlefield. In 1978 plans to develop the M113 as an IFV were finally dropped and in the following year the XM2/3 passed the Army Systems Requisition Review Council. On 1 February 1980 procurement for service production was approved by the Secretary of Defense and in October 1981 the vehicle was formally adopted and named as the M2/M3 Bradley Fighting Vehicle, after General Omar Bradley, the hero of the Normandy campaign, who had died earlier that year.

The deployment of the Bradley and Abrams in the 1980s seemed to consolidate further the preference for tracked AFVs that had emerged in the US Army at the end of World War II. The end of the Cold War did not immediately shake that mindset, but in 1999 Task Force Hawk, a combined arms force equipped with Abrams, Bradleys and other heavy equipment, was deployed from their bases in Germany to help with the NATO mission in Kosovo. It proved something of a logistical disaster, prompting the then American ambassador to the United Nations, Madeline Albright, to ask "What's the use of having the world's best military when you don't get to use them?" With the rise of a new international security situation, it was clear that the US Army's heavy forces were insufficient to meet the fast-changing demands of future combat. In other words, the US Army needed to be able to deploy quickly, by air, and make its presence felt both militarily and politically at short notice. This new strategic imperative demanded a new type of AFV: the Stryker Interim Armored Vehicle (IAV).

DEVELOPMENT OF THE STRYKER: ARMY TRANSFORMATION AND THE INTERIM FORCE

On 22 June 1999 General Eric K. Shenseki was appointed the 34th Chief of the Staff of the US Army. Shenseki had begun his career as an artillery observer in Vietnam (where he had won the Purple Heart after stepping on a landmine) and then progressed to the Armored Cavalry, serving in Europe and in the United States, eventually commanding the 1st Cavalry Division before going on to act as Commander, Land Forces Central Europe, and leading NATO's Stabilization Force (SFOR) in Bosnia. Shinseki was thus acutely aware of the changes and challenges that the US Army faced as a result of the end of the Cold War and on his appointment as Chief of Staff he outlined a bold vision for the transformation of the US Army.

Shinseki's transformation called for the establishment of an 'Objective Force' that would replace the 'Legacy Force' with its heavy, armoured vehicles that had been designed to meet the challenges of the Cold War. By contrast, the 'Objective Force' would have the same lethality and survivability as the current US Army combat formations but it would also have the flexibility to deploy anywhere in the world at short notice. At the heart of Shinseki's vision was the ability to deploy a brigade-sized force within 96 hours and a division with 120 hours. There were two vital elements to this transformation. The first was a fundamental restructuring of the US Army into Brigade Units of Action and Division Units of Employment. From henceforth, the brigade combat team was to be the principal operational unit, self-sufficient and able to fight effectively and independently. This process of transformation had been made possible in part by the changes of the previous decade, the so-called 'Forces XXI' programme. Central to this was the integration of new Command, Control, Communication and Computer (C4) technology into the Army's warfighting platforms, especially the M1 Abrams Main Battle Tank and the Bradley Fighting Vehicles. These new technologies had been developed during the 1990s and were already in place to support the new brigade combat teams of the planned Objective Force.

The second element of the new Objective Force was the replacement of the Army's current armoured fighting vehicles with a single Future Combat System (FCS). This would have consisted of a series of manned and unmanned vehicles, all linked with a sophisticated C4 apparatus. Development began in 2001 with a full design and development contract worth $14.92 million awarded two years later. It was expected that the first brigade combat team equipped with the FCS would be operational by 2015 with up to thirty brigade combat teams being so equipped by 2030. In the meantime, the Department of Defense decided to establish an 'Interim Force' to span the gulf between the Armored Brigade Combat Teams (ABCT) and more lightly equipped infantry formations. These new units required the same capabilities as the proposed Objective Force, especially the ability to deploy anywhere in the world at short notice, so the Army began the search for an armoured fighting vehicle that could help meet those requirements. This vehicle was initially known as the Medium Armored Vehicle (MAV), soon changed to the Interim Armored Vehicle (IAV).

THE INTERIM ARMORED VEHICLE

The US Army had, as we have seen, largely abandoned wheeled armoured vehicles in favour of tracked ones during the Cold War. By the late 1970s, however, several NATO countries had developed proven wheeled AFVs and in 1981 the US had decided to begin procuring a number of wheeled AFVs to give more flexibility in its response to an evolving and increasingly complex global security situation. Rather than developing their own vehicle, the US Army narrowed down their search to the British Alvis Stormer light tracked vehicle, Cadillac Gage's V-150 four-wheeled and V-300 six-wheeled armoured cars, and General Motors of Canada's eight-wheeled Piranha Light Armored Vehicle. The Piranha was itself a development of the Swiss Motorwagenfabrik AG (MOWAG) armoured car which had been developed in 6x6, 8x8 and 10x10 variants. Both the 6x6 and 8x8 versions were already in service with the Canadian army as the Grizzly APC, Cougar Fire Support Vehicle and Puma Recovery Vehicle (6X6), as well as the 8x8 Bison. The Piranha was selected from this competition but in 1984 Congress removed funding when the Army decided that the HMMWV Light Tactical Vehicle would fulfil its requirements, leaving the Piranha programme to be continued by the United States Marine Corps where it entered service in 1983 as the LAV-25. The LAV-25 went on to prove itself a versatile and successful weapons system in precisely the

(left) General Eric K. Shinseki, the US Army's 34th Chief of Staff, architect of the 'Interim Force' and the Stryker Interim Armored Vehicle. (US Army)

(above) A Canadian LAV III serving with the United Nations in Eritrea in 2001. The fact that the LAV III was already in service with the Canadian Army and had gained a reputation for mechanical reliability was one of the key factors in the choice of it as the basis for the IAV. (United Nations photo by Jorge Aramburu)

kind of contingency operations in Panama and Haiti that the proposed IAV was envisaged to take part in, as well as in the first Gulf War. The same vehicle was also adopted by the Australian Army in 1995.

Meanwhile the MOWAG Piranha continued to gain a reputation as an effective AFV. The Piranha III, especially in its 8x8 configuration, was recognised as a cost-effective and efficient AFV. In 1997 the Danish government had become the first customer of the new 8x8 Piranha IIIC. Two years later the Irish government placed an initial order for forty Piranha III, the vehicle emerging as the winner of competitive trials against the Austrian-built Steyr Pandur, while in December 2001 the Piranha III was purchased by the Spanish government for their Marines. The most important user of the MOWAG Piranha family, however, was Canada. In 1995 General Motors Diesel Division acquired a license to produce the Piranha IIIC in Canada as the Light Armoured Vehicle (LAV) III. The LAV III entered Canadian service in 1999.

The decision to procure an AFV to equip the new Interim Force was taken very quickly and in December 1999 the US Army's Tank Automotive and Armament Command (TACOM) invited eleven manufacturers to bring vehicles to Fort Knox, Texas, for a Platform Performance Demonstration. The object of this exercise was to test and review existing technology and capabilities in a series of evaluations that would last until June the following year. The manufacturers included General Dynamics Land Systems (GDLS), General Motors Defense, United Defense (manufacturer of the Bradley Fighting Vehicle) and MOWAG, as well as firms from Germany, France, Singapore and Turkey, and in all no fewer than 33 vehicles types were evaluated. They included both wheeled and tracked vehicles, prototypes such as the M8 Armored Gun System and XM1108 (based on the Bradley Fighting Vehicle chassis), as well as in-service vehicles including the Piranha IIIC 8x8, LAV III and TPz I Fuchs. On 30 December 1999 TACOM issued a Draft Request for Proposal outlining five key features of the new MAV/IAV. It was to be air-portable by a C-130 aircraft; be capable of fighting and moving in all weathers by night and day; provide front, rear and side protection against 7.62mm armour-piercing rounds; provide overhead airburst protection against 152mm rounds; and accelerate from 0-20mph on hard surfaces in no more than eight seconds.

These specifications were updated the following February with a final Operational Requirement Document. The IAV would need to be configured in a number of roles, principally as an Infantry Carrier Vehicle (ICV) capable of carrying a full infantry squad, but also as a reconnaissance vehicle, mortar carrier, engineer squad vehicle, command post vehicle, fire support vehicle, anti-tank guided missile platform, ambulance and Nuclear, Biological and Chemical (NBC) warfare vehicle. Future variants might include a Mobile Gun System and a 155mm self-propelled howitzer variant. In the same month GDLS and General Motors Defense of Canada announced they had joined forces to offer the LAV III as the basis for the proposed IAV. The Canadian Army offered 32 LAV III for evaluation and training to the US Army, while sixteen Centauro 8x8 vehicles armed with a 105mm gun arrived from Italy. For the next year these vehicles were extensively tested to evaluate the tactical doctrine for the new Interim Brigade Combat Teams.

On 16 November 2000 TACOM awarded a contract worth some $4 billion to the GM GDLS Defense Group with a six-year contract to supply 2,131 IAV to equip the US Army's new IBCT. Initially, 366 vehicles

(above) Air portability was one of the key operational requirements for the IAV. Here a 2nd Cavalry Regiment Stryker off-loads from a C-17 Globemaster III at Ramstein Air Base, Germany, in support of Exercise Steadfast Javelin II in September 2014. (US Air Force photo by Senior Airman Damon Kasberg)

in eight different variants were to be produced at a cost of $578.4 million. The decision to award the contract to the GM GDLS Defense Group was controversial, with one of the other manufacturers involved, United Defense, protesting that the procurement process had not been conducted in a fair and transparent way. The US Government's General Accounting Office ordered work on the IAV programme to stop and held an investigation. They concluded that the cost benefits of the LAV III and the support offered by GM GDLS Defense Group in maintaining the fleet outweighed the advantages of United Defense's Medium Tactical Vehicle Light (based on the M113 chassis) and M8 Armored Gun System. Eventually, in April 2001 work recommenced on the IAV programme. Low-rate production began later that month and the first deliveries to the US Army were made from General Motors' plant in Ontario in March 2002, followed a month later by the first vehicles from GDLS's plant at Anniston, Alabama.

On 27 February 2002 in a ceremony at Fort Lauderdale, Florida, the IAV was christened the Stryker. The naming was unusual as it was the first time that a US AFV had been named after enlisted men rather than a general. The name remembered two Medal of Honor winners: Pfc. Stuart Stryker of the 513th Parachute Infantry had been killed in March 1945 during Operation Varsity, leading an attack on German positions which had resulted in the capture of 200 enemy and the freeing of three captured US airmen; and Spc. Robert Stryker of the 1st Infantry Division who had posthumously received the award for saving the lives of several of several wounded fellow soldiers by taking the blast from a Claymore mine in Loc Ninh, Vietnam, in November 1967.

(left) An M1128 Mobile Gun System and M1126 Infantry Carrier Vehicle of 2-23 Infantry, 4th Stryker Brigade Combat Team, 2nd Infantry Division, at the Yakima Training Center, WA, in October 2012. (US Army photo by Spc. Kimberley Hackbarth)

(right) A squad dismounts from an M1126 ICV of 3-21 Infantry, 1st Brigade, 25th Infantry Division in Mosul, Iraq, in May 2005. Note the slat armour and the necessary addition of Kevlar rolls and sandbags to the roof of the Stryker. (US Army photo by Spc. Jory C. Randall)

THE STRYKER: AN EFFECTIVE WEAPON

The production of a further 212 Strykers was ordered on 4 March 2004, following positive reports of its first employment in Iraq. A further 116 vehicles were ordered in June and in October the Department of Defense (DoD) approved low-rate production of the 105mm cannon-armed M1128 Mobile Gun System and the M1135 NBCRV. In December the DoD placed a further order for 95 Strykers. In January the 1000th Stryker rolled off the production line at the Anniston Army Depot in Alabama, with some 800 in service with the three then operational Stryker Brigade Combat Teams (SBCT). The reputation of the vehicle was quickly growing: the first SBCT to be deployed to Iraq had done so with 311 Strykers and they had maintained a 97% rate of operational availability throughout their tour. In February 2005 the DoD put in another massive order, worth $582 million, for 423 more vehicles to equip a fifth SBCT, quickly followed by another order for 99 Strykers, at a cost of $138 million, in April.

By the close of 2005, the Strykers in Iraq (shared between the two SBCT deployed there) had travelled over six million miles and been in continuous combat for more than two years. In November TACOM ordered an overhaul of 265 vehicles to like-new condition in a contract worth $69 million. In February the following GDLS established a repair facility in Qatar to return damaged vehicles to the frontline in Iraq. The Stryker was proving its worth and proving popular with its crews. In April 2006 306 Strykers, worth $464 million were ordered to begin equipping three new SBCT. In June and July further orders for 206 vehicles at a cost of $254 million were placed. In all 2,691 Stryker IAV were ordered to equip the new SBCT between 2000 and 2006.

IMPROVING THE STRYKER

Unsurprisingly, as a vehicle borne out of an urgent operational requirement and quickly thrust into the crucible of war, the Stryker family of vehicles has undergone a number of changes during its service history. The basic Stryker is 6.98m long, 2.72m wide and 2.64m high (excluding any armament). It has 85 per cent commonalty of parts across its ten variants. The ICV version had an initial combat weight of 18,600kg. It is powered by a diesel Caterpillar 3126 engine which gave the first vehicles a top speed of 96.5kph. It had eight Michelin 1200R20 XML tyres that had a run-flat capability. The vehicle is not amphibious, but can ford a depth of 1.06m without preparation. The vehicle's main armour was 1/2in high-hardness steel capable of defeating 7.62mm rounds, but production vehicles also had additional 3mm steel and ceramic plates that provide protection against 14.5mm rounds.

The first modification made to the Stryker was the fitting of a set of slat armour to vehicles serving in Iraq. These were designed to detonate shaped charges, such as those fired from handheld rocket launchers like the RPG-7, before they impacted on the vehicle's main armour. The slat armour kit made by GDLS weighs 2,361kg and development began before the Stryker deployed to Iraq. In action, while the slat armour successfully defeated High Explosive Anti-Tank (HEAT) RPG rounds, it proved less successful against Anti-Personnel (AP) and Anti-Tank (AT) rounds. Shrapnel from the former or the penetrator of the latter could bypass the slats and impact on the vehicle's surface, leading to crews piling sandbags and Kevlar rolls on their Strykers which further compromised the vehicle's performance, already adversely affected by the extra weight of the slat armour. As a result, in March 2005, BAE Land Systems were awarded a contract to fit further add-on armour to 289 Strykers. This increased the vehicle weight by 3,100kg.

Other changes were introduced throughout the initial production batches and retrofitted to existing vehicles. In 2004 the Army approved the upgrade of the M151 Remote Weapons Station (RWS) to M151E1 standard. This included a new Thermal Imaging System and Video Imaging Module and was introduced on Strykers for the fifth SBCT to be activated. This was followed by a M151E2 improvement, which introduced a stabilisation for the RWS for speeds up to 40km/h and is now standard for the Stryker family. One of the biggest threats to the Stryker in Iraq was the Improvised Explosive Device (IED). To combat this threat the Driver's Enhancement Kit (DEK), an additional armour package fitted to the lower hull and designed to protect the vulnerable driver's compartment, as introduced. In March 2009 a contract for 805 DEKs worth $4.8million was awarded to GDLS – Canada. Other improvements included a Driver's Vision Enhancer (DVE) fitted to the right of the driver's hatch and a foldable windshield kit which allows the Stryker to drive in poor conditions with the hatch open. It can be stowed in the vehicle when not in use.

Other changes have mostly been digital and designed to improve the Strykers' C4 capabilities and situational awareness. Around 2014 a new FBCB2/BFT Joint Capabilities Release (JCR) with a new Blue Force Tracker (BFT-2) antenna was introduced, allowing updates of friendly forces' positions within seconds of being refreshed.

DOUBLE-V HULL STRYKER
In 2009 the much-vaunted ground Future Combat System was cancelled (at a cost of over $1 billion). In its place the Stryker programme continued to develop. The most important production change introduced to the Stryker is the Double-V Hull. As early as 2007, as a result of the combat experience of the Stryker and LAV III in Iraq and Afghanistan, the need for additional hull protection was recognised, but it was not until July 2010 that GDLS was awarded a $30 million contract to begin production of a new hull. The Double-V Hull (DVH) is actually a 'w'-shaped bottom to the hull, designed to deflect the blast away from the crew compartment. The Double-V Hull package is much more than simply reconfiguring the shape of the hull: it also includes mine-resistant blast seating for the squad compartment, a new suspension system, wider tyres, a height management system and other additional protection. The development of the DVH was not straightforward and it was not until July 2011 that the first 450 DVH Strykers were ordered into production. This was quickly followed by a second order for 742 vehicles and a further 760 in 2012. The first two SBCT had converted to DVH by January 2014 and the process of conversion is still on-going. Not all Stryker variants have been produced as DVH vehicles. The M1127 RV, M1128 MGS and M1135 NBCRV are only seen in the flat-bottomed variants.

Concurrent with the DVH programme is the Stryker Engineering Change Proposal (ECP) programme. These changes are mainly internal and include a more powerful engine, improved suspension, and greater on-board electrical power to facilitate further digital improvements designed to enhance situational awareness. These vehicles with the upgrades to the automotive and electrical systems, as well as the DVH, are known as Stryker A1. In June 2018 the Army awarded GDLS a $258 million contract to upgrade its remaining flat-bottom Strykers to the A1 standard by the spring of 2020.

In all some 4,900 Strykers have been built and 4,466 serve with the US Army. Despite its detractors – who argued that it was too costly, poorly protected and not fit to meet the rigours of constant war in the years after 9/11 – the Stryker has proved itself an effective and popular platform that has served the men and women of the US Army well in peacetime and in war. As we shall, the versatility and reliability of the basic design remain one of its key advantages and the principal reason why it is likely to remain in the US Army's inventory for some time to come.

(left) Strykers in the field are constantly being modified at unit level. Note the extended jerry can holders fitted to the Stryker ICV-J of 3-2 Cavalry photographed at the Bemowo Piskie Training Ground, Poland, in May 2020. (US Army photo by Sgt. Timothy Hamlin)

(above) An M1126 OF 1st SBCT, 25th Infantry Division, during training at the Joint Readiness Training Center, Fort Polk, LA, in March 2004. (USAF photo by SRA Jorge A Rodriguez)

THE STRYKER BRIGADE COMBAT TEAM

The adoption of the Stryker in the US Army in 2002 marked not only the introduction of a new armoured fighting vehicle but also a new concept in the American ability to fight wars. As we have seen, the new US Army Chief of Staff, General Eric K. Shinseki, had demanded a 'interim force' that could deploy quickly to the world's troublespots and protect American interests as part of his programme to transform the US Army to meet the challenges of the new century. In 2000 the Army issued its Operational Requirements Document (ORD, 2000) for the new Interim Armored Vehicle and defined the top-level requirements for the new Interim Brigade Combat Team (IBCT):

'The IBCT is a full spectrum, combat force. It has utility … in all operational environments against all projected threats, but it is designed and optimised primarily for employment in small scale contingency (SSC) operations in complex and urban terrain, confronting low-end and mid-range threat that may employ both conventional and asymmetric capabilities. The IBCT deploys very rapidly, executes early entry, and conducts effective combat operations immediately on arrival to prevent, contain, stabilise, or resolve a conflict through shaping and decisive operations … As a full spectrum combat force, the IBCT is capable of conducting all major doctrinal operations, including offensive, defensive, stability, and support operations … Properly integrated through a mobile robust C4ISR network, these core capabilities compensate for platform limitations that may exist in the close fight, leading to enhanced force effectiveness.'

The IBCT was designed to be capable of being deployed anywhere in the world within 96 hours and once its entry point had been secured by paratroopers or Special Forces, the brigade should be able to conduct independent operations for up to 72 hours. While its original planners had seen the IBCT as optimised for SSC operations, the first Field Manual for the new formation, issued on 13 March 2003 (FM.3-21.31), made it clear that it was expected to operate just as effectively in Major Theatre War (MTW) and Peacetime Military Engagements (PME). At the heart of the newly named Stryker Brigade Combat Teams' (SBCT) warfighting capabilities was an 'enhanced situational understanding'. The SBCT employed a multi-level, integrated suite of intelligence, reconnaissance and surveillance assets to deliver a common operational picture shared by all elements of the force. Both digital systems and human intelligence (HUMINT) combined to allow the SBCT commander what the planners called 'full-spectrum flexibility', allowing him to plan and execute several missions simultaneously, deploying different elements of the SBCT in different places while maintaining an overall operational picture. This, as we will see, was key to the SBCT's effectiveness in Iraq and Afghanistan.

From the beginning, then, the SBCT team was organised to meet the requirement for meeting the full range of missions the US Army might be called upon to perform. Initially the SBCT consisted of some 3,500 soldiers and while this number has increased to nearer 4,500 in recent years the fundamental organisation of the unit remains unchanged. At the heart of the SBCT are the three Stryker infantry battalions. These are organised in a 'three-by-three' structure of three companies, each with three platoons. According to the 2012 US Army Force Structure Data Manual (Supplemental Manual 3-90) each rifle company has fourteen Infantry Carrier Vehicles (ICV), two Mortar Carrier Vehicles (MCV) and three Mobile Gun Systems (MGS). Each battalion also has a Headquarters company consisting of five Command Vehicles (CV), three Fire Support Vehicles (FSV), four Medical Evacuation Vehicles (MEV), four MCV and four Reconnaissance Vehicles (RV). It also has over twenty HMMWV trucks in various configurations.

Central to the SBCT's ability to conduct full-spectrum missions is the Reconnaissance, Surveillance and Target Acquisition Squadron (RSTA). As it was originally conceived, the SBCT is highly dependent on intelligence, surveillance and reconnaissance to mitigate any shortcomings in the Stryker platform and the brigade's lack of firepower compared to its Legacy Force counterparts. The RSTA's squadron's primary role is to provide accurate and timely intelligence, acting as the 'eyes and ears' of the brigade commander, allowing him to concentrate the brigade's combat power at the decisive time and place. As one of the initial assessments of the IBCT structure observed: 'the IBCT gains its lethality and survivability from maneuver and maintaining positional advantage over an opponent ... Military intelligence is a major contributor to achieving this capability'. Under the 2012 organisation, the RSTA squadron is comprised of a Headquarters troop with six CV, three FSV and four MEV, as well as eighteen HMMWV. Each of the three reconnaissance troops is equipped with a CV, two MCV and thirteen RV, as well as a HMMWV.

Direct fire support is provided to the SBCT by its Fires or Field Artillery Battalion. This consists of three batteries, originally equipped with six M-198 155mm howitzer but now fielding the M777 155mm Light Towed Howitzer. The battalion only includes one CV to maintain seamless communications and situational awareness with the brigade commander, but has over fifty HMWVV and a similar number of Medium Tactical Vehicles (MTV) to tow the howitzers into action and perform other support tasks. Each SBCT also has a separate anti-armour company. This was equipped with the Anti-Tank Guided Missile Vehicle (ATGMV) and attached to the Brigade Engineer Battalion, but in 2015 it was moved to the RSTA squadron and now incorporates the MGS into its order of battle.

The Headquarters of the SBCT contains everything necessary to plan, direct, control, provide management and communications, and to coordinate the brigade's operations. It is organised into a Headquarters company equipped with four CV, four FSV (equipped with M707 Stryker Mission Equipment Package) and six M1117 Armoured Security Vehicles, as well as over fifty HMMWV and MTV. The SBCT also has a Brigade Engineer Battalion, consisting of the Signal Company with two CV, the Military Intelligence Company, a Tactical Unmanned Aircraft System (TUAS) platoon, and two Combat Engineer Companies with twelve Engineer Squad Vehicles (ESV) and one ICV each. The SBCT is also supported by the Brigade Support Battalion, which

(below) Commanche Troop, 3-2 Cavalry during a mission rehearsal exercise at the Joint Multinational Readiness Center in Hohenfels, Germany in March 2013. (US Army photo by Spc. Jordan Fuller)

(above) Direct fire support is offered to the SBCT by the M777 155mm Light Towed Howitzer. Here troopers of Archer Battery, Field Artillery Squadron, 2nd Cavalry Regiment, conduct a live-fire exercise at the Grafenwoehr Training Area in February 2015. (US Army photo by Nathanael Mercado)

contains field kitchens, a transport and distribution company, a field maintenance company, and a medical company.

In December 2003 the DoD announced funding to convert six brigades to SBCT, an increase on the two IBCTs the Army had originally envisaged back in 2000. The first unit to convert to a SBCT was the 3rd Brigade, 2nd Infantry Division, based in Fort Lewis Washington. This, as we shall see, was also the first unit to deploy operationally with the Stryker when it arrived in Iraq in November 2003. In April 2003 conversion of a second unit, 1st Brigade, 25th Infantry Division, began at Fort Lewis. This was the second SBCT to deploy to combat operations in Iraq. In June 2006 it was reflagged as 2nd Stryker Cavalry Regiment (later simply known as 2nd Cavalry Regiment) and moved to Rose Barracks in Vilseck, Germany, where it remains as the only SBCT permanently assigned to USAEUR (United States Army Europe).

The third SBCT to be activated was 172nd Infantry Brigade in July 2004, based in Fort Wainwright, Alaska. On its return from Iraq in November 2006 it was reflagged as 1st Brigade, 25th Infantry Division. The fourth SBCT to be activated was 2nd Armored Cavalry Regiment which was converted to Strykers and had its name changed to 2nd Stryker Cavalry Regiment following its return from Iraq in April 2005. In June 2006 it was reflagged as 4th Brigade, 2nd Infantry Division and re-homed to Fort Lewis, Washington. In late 2005 a fifth brigade, 2nd Brigade, 25th Infantry Division, based at Schofield Barracks, Hawaii, began conversion to a SBCT, and in 2007 the 5th Brigade, 2nd Infantry Division also began to be equipped with Strykers. In July 2010 this latter unit was deactivated and reflagged as 2nd Brigade, 2nd Infantry Division.

The number and designation of SBCT within the US Army is constantly under review. Following its return from Iraq in November 2010, 1st Brigade, 1st Armored Division, based at Fort Bliss, Texas, surrendered its Abrams MBT and Bradley Fighting Vehicles and began conversion to a SBCT. As such it deployed to Afghanistan in December 2012 and again four years later. On 16 November 2011 the 3rd Armored Cavalry Regiment, based at Fort Hood, TX, also converted from an Armored Brigade Combat Teams (ABCT) to a SBCT and deployed to Afghanistan in June 2014. On 20 June 2019, however, in response to the Army's need for more ABCT, the 1st SBCT, 1st Armored Division, gave up its Strykers in favour, once again, of MBTs and Bradleys. At the same 2nd Brigade, 4th Infantry Division began the conversion from an Infantry Brigade Combat Team (IBCT) to a SBCT. Currently there are seven SBCT in the US Army, alongside eleven ABCT and thirteen IBCT.

There are also two United States National Guard SBCT. The 56th Brigade, 28th Infantry Division, Pennsylvania Army National Guard began its conversion to Strykers in October 2004, receiving its first vehicles seven months later. In January 2009, after four months of intensive training, it became the first National Guard Stryker unit to deploy to Iraq. In July 2015 the 81st Brigade, 7th Infantry Division, converted from an ABCT to a SBCT. The brigade contains elements from the California, Oregon and Washington National Guards.

The SBCT, like the Stryker itself, has proved itself an effective and versatile part of the United States' warfighting capability. Its role, and the threats it is designed to face, have changed greatly since its inception in 2000. In 2014 the Russian annexation of Crimea and the civil war in Ukraine raised once again the spectre of conflict between NATO and Russia. As a result, the US Army rushed to reorientate its forces towards war with a near-peer adversary. The suitability of the SBCT to meet this new threat was soon called into question. Critics point out that it was, essentially, an infantry formation designed to carry soldiers safely into battle, unlike the Bradley Fighting Vehicle which was designed to fight through to the objective alongside the Abrams MBT. Two programmes, as we shall see, were designed to increase the SBCT's lethality and survivability against a near-peer adversary: the 30mm cannon and Javelin ATGM mounted on an ICV. Nevertheless, the Stryker is not sufficiently armoured to engage enemy armour (or even enemy infantry fighting vehicles like the BMP-3). Another shortcoming of the SBCT is the lack of organic Short-Range Air Defence (SHORAD). This, of course, had not been an issue in the type of SSC or counter-insurgency operations the Stryker was initially designed for and has been deployed for in Afghanistan and Iraq. Against a near-peer adversary, however, the lack of SHORAD could prove disastrous for the SBCT and the provision of such assets within the SBCT is currently one of the US Army's top priorities. It is hoped to deploy Stryker-based SHORAD operationally by the end of 2020.

Regardless of the technical shortcomings of and developments to the Stryker family of vehicles, the key asset of the SBCT is the men and women who serve in them. A rigorous programme of training, the trust shown in and independence given to junior officers and NCOs, and the invaluable combat experience gained in Afghanistan and Iraq, will ensure that the SBCT remains an integral part of the US Army in the decade to come and beyond.

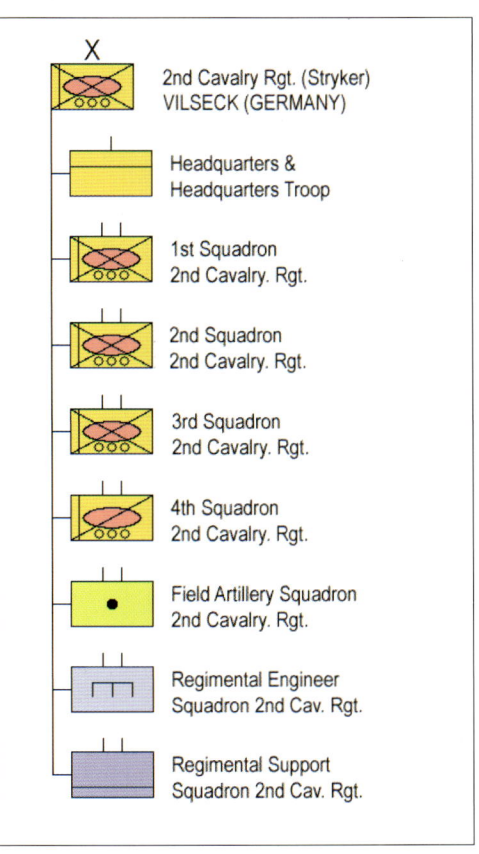

(left) The 2nd and 3rd Cavalry Regiments are independent regiments outside of the US Army's divisional system and retain their cavalry lineage in their organisation as SBCT.

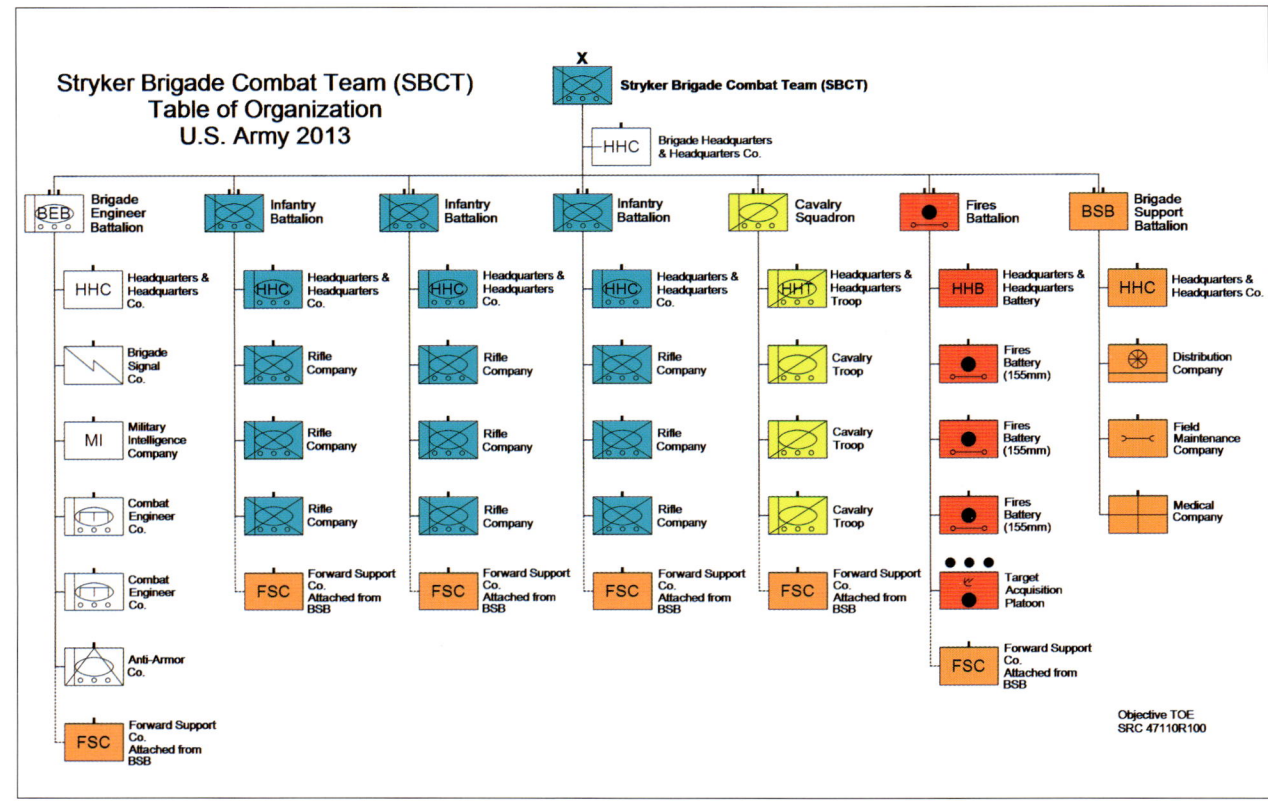

(below) The standard table of organisation of an US Army Stryker Brigade Combat Team (CrucibleX).

(top) This image gives a good sense of how the M1126 ICV functions as a 'battle taxi' for the infantry. Here a section from 1-2 Cavalry deploys during Exercise Saber Junction in 2012. (US Army photo by Spc. Ian Schell)

M1126 STRYKER INFANTRY CARRIER VEHICLE (ICV)

The M1126 is the standard Stryker variant, forming the bulk of the Stryker Brigade Combat Team (SBCT), with each having 108 ICVs. It carries a nine-man infantry squad, alongside the two-man vehicle crew. It is equipped with the M151 Protector Remote Weapon Station (RWS), armed with either the .50cal M2HB Machine Gun or the 40mm Mk19 Mod 3 Automatic Grenade Launcher. 2,000 rounds are carried for the .50cal (or 430 rounds for the 40mm grenade launcher). The ICV also carries a M240 7.62mm machine gun that can be mounted externally on the commander's hatch. As well as 3,200 rounds of 7.62mm ammunition, the ICV also carries four Javelin Anti-Tank Guided Missiles (ATGMs).

Even before it entered service reservations were expressed about the ICV. It is important to note that the ICV is not a Infantry Vehicle Fighting (IFV) like the M2 Bradley, but primarily a means of armoured transportation for the infantry elements of the SBCT. The infantry section cannot see outside the vehicle, let alone fire their personal weapons, although they do have access to the Video Display Terminal linked to the RWS and the squad leader can use the driver's enhanced imaging camera.

A more important consideration is the inability of the basic ICV to engage successfully a peer or near-peer opponent, especially one equipped with armour like the BMP-3 IFV used by the Russian Federation. This lack of firepower was, as we shall see, already evident in the

(right) A good view of the Mk19 40mm grenade launcher fitted to the RWS of a M1126 of Eagle Troop, 2-2 Cavalry. (US Army photo by Spc. Derek Hamilton)

(left) Fox Troop, 2-2 Cavalry M1126 fitted with the CROWS-J (but without the Javelin missile itself) prior to live-fire trials at the Grafenwoehr Training Area in October 2018. (US Army photo by 1st Lt. Ellen C. Brabo)

Stryker's experience in Iraq, but the issue became more acute as the United States began to turn its attention away from low-intensity, counter-insurgency warfare even before the Russian intervention in Ukraine in 2014. In February that year an exercise was held at the National Training Center, Fort Irwin, CA, that pitted 3rd SBCT against a conventional combined arms force. The SBCT fared well, 'destroying' two-thirds of the armoured vehicles on the opposing side, while losing less than a third of their own vehicles. The tactics used, however, were to dismount and engage enemy armour with Javelin ATGMs. In 2015 the 1st SBCT, 1st Armored Division, engaged a similar tank-heavy force at the National Training Center. On this occasion, the shortcomings of the SBCT were more obvious and the experience led to calls for a 30mm gun and Remote Weapons Station – Javelin (RWS-J) to be installed and fielded 'at the earliest opportunity'.

Calls to provide the Stryker with the means to engage enemy AFVs were led by the 2nd Cavalry Regiment, based in Germany, and saw the introduction of the M1296 Stryker Dragoon ICV. Another innovation by 2nd Cavalry Regiment was to equip certain ICVs with the Stryker Common Remotely Operated Weapon Station – Javelin (CROWS-J). This was in response to an Operational Needs Statement submitted by 2nd Cavalry Regiment in March 2015 and aimed to equip 81 ICVs, 50 per cent of the rifle and scout platoons, with RWS-mounted Javelins alongside their .50cals or 40mm grenade launchers. Early field trials in the spring of 2018 in Germany confirmed the efficacy of the system in both live-fire scenarios (with five out of six hits) and in a tactical exercise against an armoured opponent where the Strykers successfully completed fourteen of the sixteen assigned missions. Nevertheless, the trials also identified cyber security issues, as well as problems with the mount of the Javelin system on the existing RWS. Efforts to develop a different system in the United States had also stalled by the autumn of 2019. It is hoped to fully deploy the CROWS-J with 2nd Cavalry in the summer of 2020 and then equip the US-based Stryker brigades in stages up to 2022.

(left) A close up of the problematic CROWS-J developed for 2nd Cavalry's Strykers. (US Army photo by Markus Rauchenberger)

(above) An FV of 4-2 Cavalry patrols the streets of Baghdad in September 2007. Note the slat armour and the RV/FSV Commander's Cupola Shield. (US Army)

(below) Even the M1127 RV carries a large amount of stowage as shown by this 4-2 Cavalry vehicle during Exercise Allied Spirit I at the Joint Multinational Readiness Center in Hohenfels, Germany, in January 2015. (US Army photo by Pfc. Lloyd Villanueva)

M1127 STRYKER RECONNAISSANCE VEHICLE (RV)

Reconnaissance has been, and continues to be, at the heart of the cavalry's role on the battlefield. In this regard, the SBCT is no different to its predecessors. The reconnaissance role is principally the task of the scout platoons of the infantry battalions and the Renaissance, Surveillance and Target Acquisition (RSTA) Squadron equipped with the M1127 Stryker Reconnaissance Vehicle (RV). As well as the two-man crew, there is a four-man reconnaissance squad and intelligence soldier in each vehicle. The M1127 mounts a Long-Range Advanced Scout Surveillance System (LRA3S) pod, alongside either a pintle-mounted M2HB or Mk. 19 grenade launcher fitted to the commander's cupola instead of the M151 RWS. 48 are issued to each SBCT. The LRA3S has the ability to detect targets up to a range of 15km. A laser-rangefinder allows the scouts to calculate distance accurately and provide a grid reference for direct fire support or other tasks. The LRA3S also features an infrared thermal imager, a day video camera, long-range common aperture reflective optics, and a GPS interferometer subsystem.

In Iraq it became clear that the four-man reconnaissance squad was insufficient to allow for effective scouting, particularly in urban areas, while also maintaining the security and operability of the M1127. Platoons would frequently keep a vehicle behind at the Forward Operating Base (FOB) to ensure sufficient 'boots on the ground' with HUMINT (intelligence gathering by dismounted soldiers) the most important aspect of the scout platoons' work. It was probably because of this that the M1127 did not receive the new Double-V Hull (DVH) improvements. Indeed, a new version of the ICV, the ICVV-S, has been developed for service in the scout platoons. This has the double hull, as well as other Engineering Change Proposal Enhancements, and is equipped with the Scout Mission Equipment Package (MEP) consisting primarily of an internally mounted LRA3S.

(left) Another 4-2 Cavalry M1127 RV, this time photographed during Exercise Slovak Shield in October 2016. (US Army photo by Staff Sgt. Micah Van Dyke)

(above) A scout platoon of 1-36 Infantry, 1st SBCT, 1st Armored Division, moves into position at the National Training Center, Fort Bliss, TX, in April 2018. (US Army photo by Winifred Brown)

(left) A good close up of the LRA3S fitted to the M1127 RV at the Grafenwoehr Training Area in May 2012. (US Army by Getrud Zach)

16 Stryker Interim Combat Vehicle

1. **M1126 ICVV-S, Palehorse Troop, 4th Squadron, 2nd Cavalry Regiment, Bemowo Piskie Training Area, Poland, September 2015.**
This Stryker, training with Polish paratroopers, is finished in the usual Bronze Green (FS34094). This Stryker serves with the 4th, RSTA Squadron of 2nd Cavalry Regiment. Instead of the M1127 RV, the Squadron has some DVH M1126 Stryker ICV with an internally mounted Long-Range Advance Scout (LRAS) Surveillance System. Note the Blue-Force Tracker, added alongside the driver's hatch, to most 2nd Cavalry Strykers on 2014. As part of the 2016 Engineer Change Proposal 1 programme, further M1126 DVH will be upgraded to ICVV-A1 standard with the scout Mission Equipment Package (MEP) and replace the M1127 in the SBCT.

(US Army photo by Spc. Marcus Floyd)

COLOUR PROFILES BY SLAWOMIR ZAJACZKOWSKI

2. M1127 RV, Mustang Troop, 1st Squadron, 2nd Cavalry Regiment, Exercise Saber Strike 18, Bemowo Piskie Training Area, Poland, June 2018. Four M1127 RV serve in the scout platoon of the Headquarters Troop of each of the Stryker squadrons of 2nd Cavalry Regiment (equivalent to the infantry battalion of the standard SBCT).

3. M1128 MGS, 2nd Cavalry Regiment, Grafenwoehr Training Area, Germany, March 2012. Finished in Bronze Green with a heavy coating of dust thrown up from the live-fire exercise.

(below) An M1127 RV of 5-1 Cavalry, 1st SBCT, 25th Infantry Division, the Yukon Training Area, Alaska, in August 2014. (US Air Force photo by Justin Connaher)

4. M1130 CV, 1st Battalion, 23rd Infantry Regiment, 3rd Stryker Brigade Combat Team, 2nd Infantry Division, Tal Afar, Iraq, September 2004. This Command Vehicle is fitted with the slat armour that was standard to vehicles in Iraq.

5. M1131 FSV, Palehorse Troop, 4th Squadron, 2nd Cavalry Regiment, Nowa Deba Training Area, Poland, July 2015. This M1131 FSV from the RSTA squadron of 2nd Cavalry Regiment is distinguishable from the M1127 RV only by the extra antennas.

(NARA: Sgt. Jeremiah Johnson)

6. M1132 ESV, 2nd Cavalry Regiment, Task Force Dragoon, Zangabad, Afghanistan, July 2013. This 2nd Cavalry Regiment Stryker is fitted with a Surface Mine Plow and an improvised camouflage net sun shelter.

7. M1135, 1st Stryker Brigade Combat Team, 1st Armored Division, White Sands Missile Range, United States of America, February, 2019.

A 2nd Cavalry Regiment ESV with a Straight Obstacle Blade during Exercise Allied Spirit IV in January 2016. (US Army photo by Sgt. William Tanner)

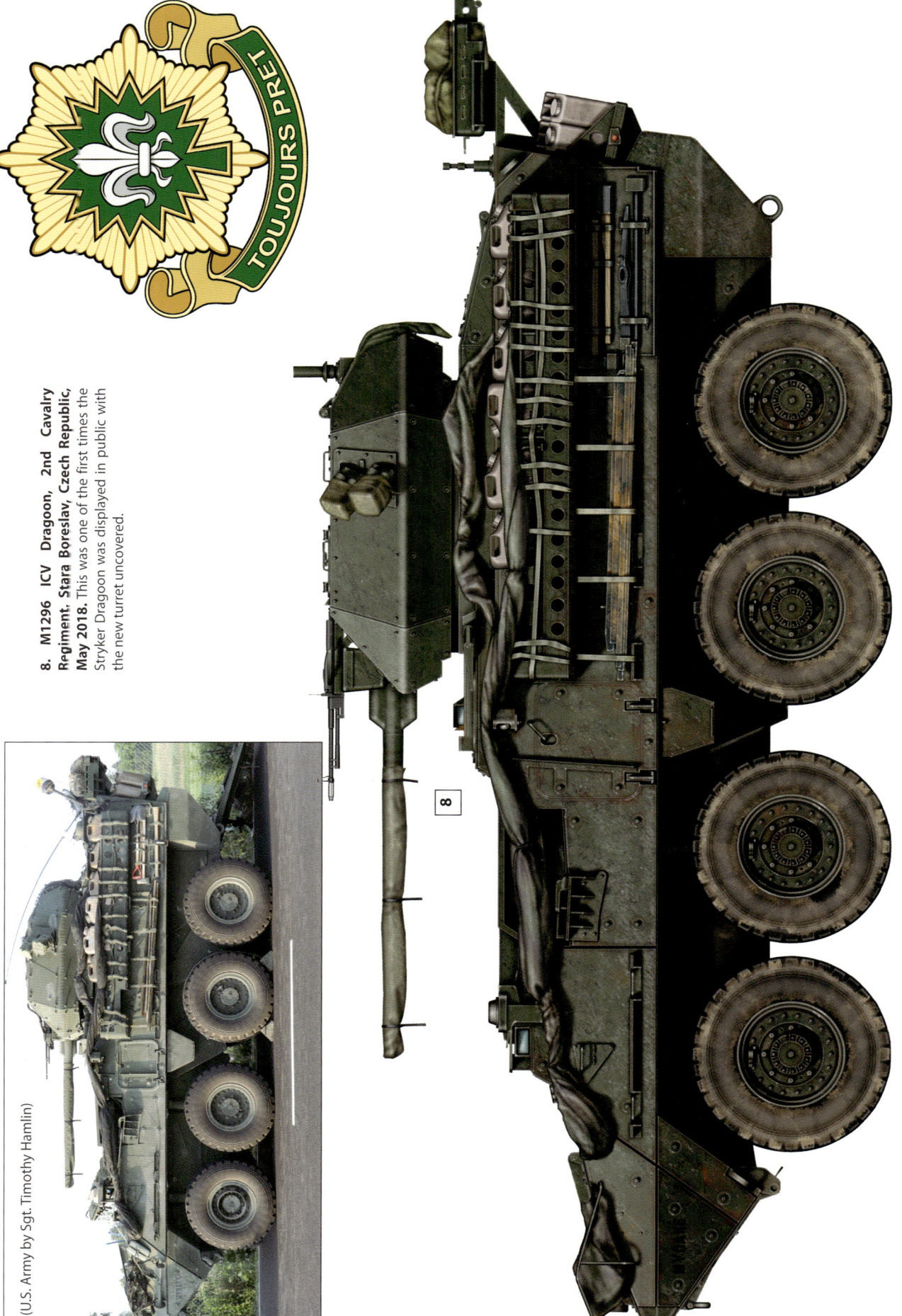

8. **M1296 ICV Dragoon, 2nd Cavalry Regiment, Stara Boreslav, Czech Republic, May 2018.** This was one of the first times the Stryker Dragoon was displayed in public with the new turret uncovered.

(U.S. Army by Sgt. Timothy Hamlin)

Camouflage and Markings 21

(ASO Network)

(ASO Network)

9. M1126 ICVV-A1, 75th Ranger Regiment, Rojava, Syria, April 2017. This double-v hull Stryker was one of a number of US vehicles filmed and photographed operating patrols in north-eastern Syria in the spring of 2017. The Strykers were operated by 75th Ranger Regiment and unusually some were painted in CARC Tan. They had a number of unique features including extra lighting and additional armour around the top of the hull.

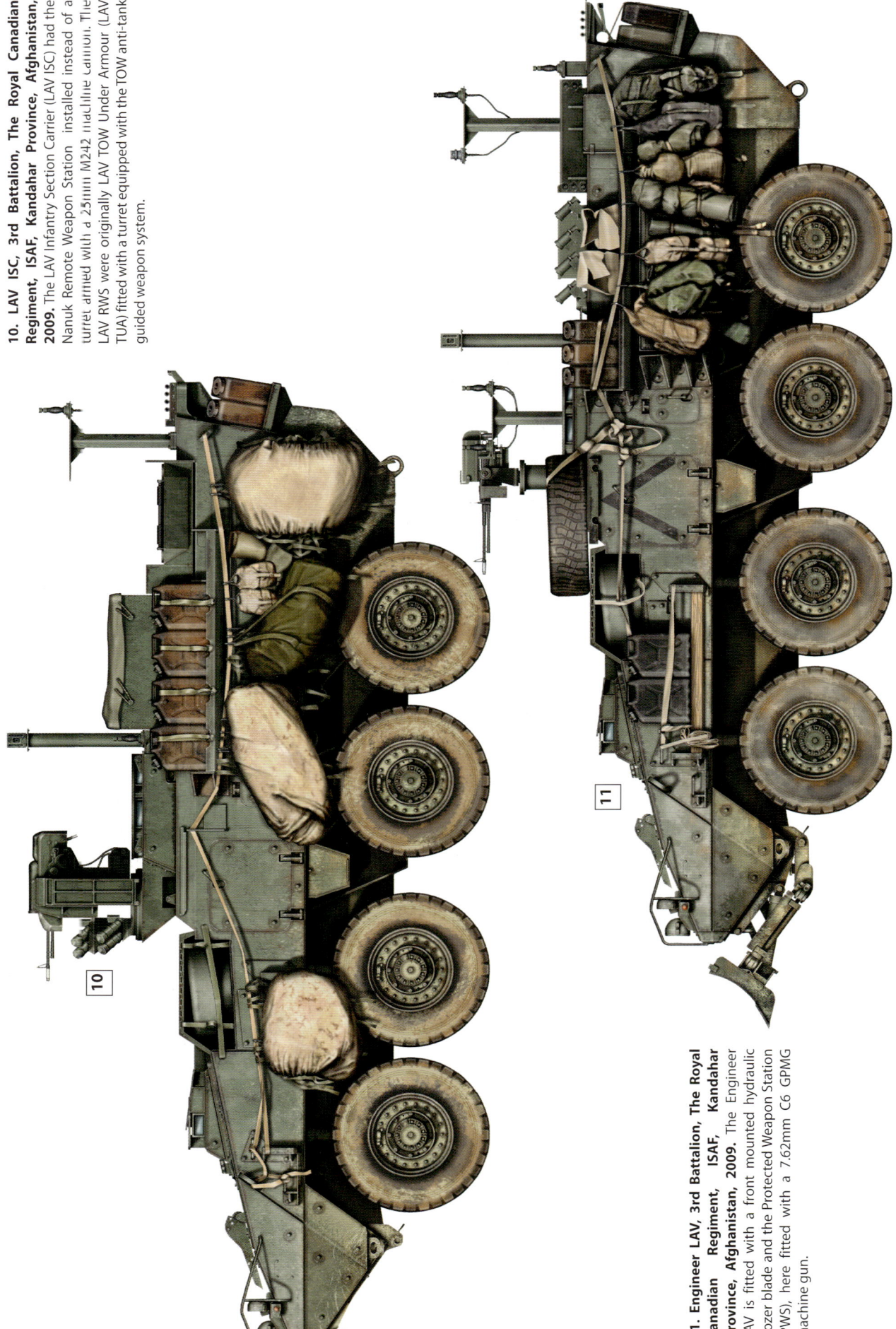

10. LAV ISC, 3rd Battalion, The Royal Canadian Regiment, ISAF, Kandahar Province, Afghanistan, 2009. The LAV Infantry Section Carrier (LAV ISC) had the Nanuk Remote Weapon Station installed instead of a turret armed with a 25mm M242 machine cannon. The LAV RWS were originally LAV TOW Under Armour (LAV TUA) fitted with a turret equipped with the TOW anti-tank guided weapon system.

11. Engineer LAV, 3rd Battalion, The Royal Canadian Regiment, ISAF, Kandahar Province, Afghanistan, 2009. The Engineer LAV is fitted with a front mounted hydraulic dozer blade and the Protected Weapon Station (PWS), here fitted with a 7.62mm C6 GPMG machine gun.

12 and 13. LAV 6.0, Royal 22e Régiment, Exercise Common Ground II, 5th Canadian Division Support Base, New Brunswick, Canada, November 2019.

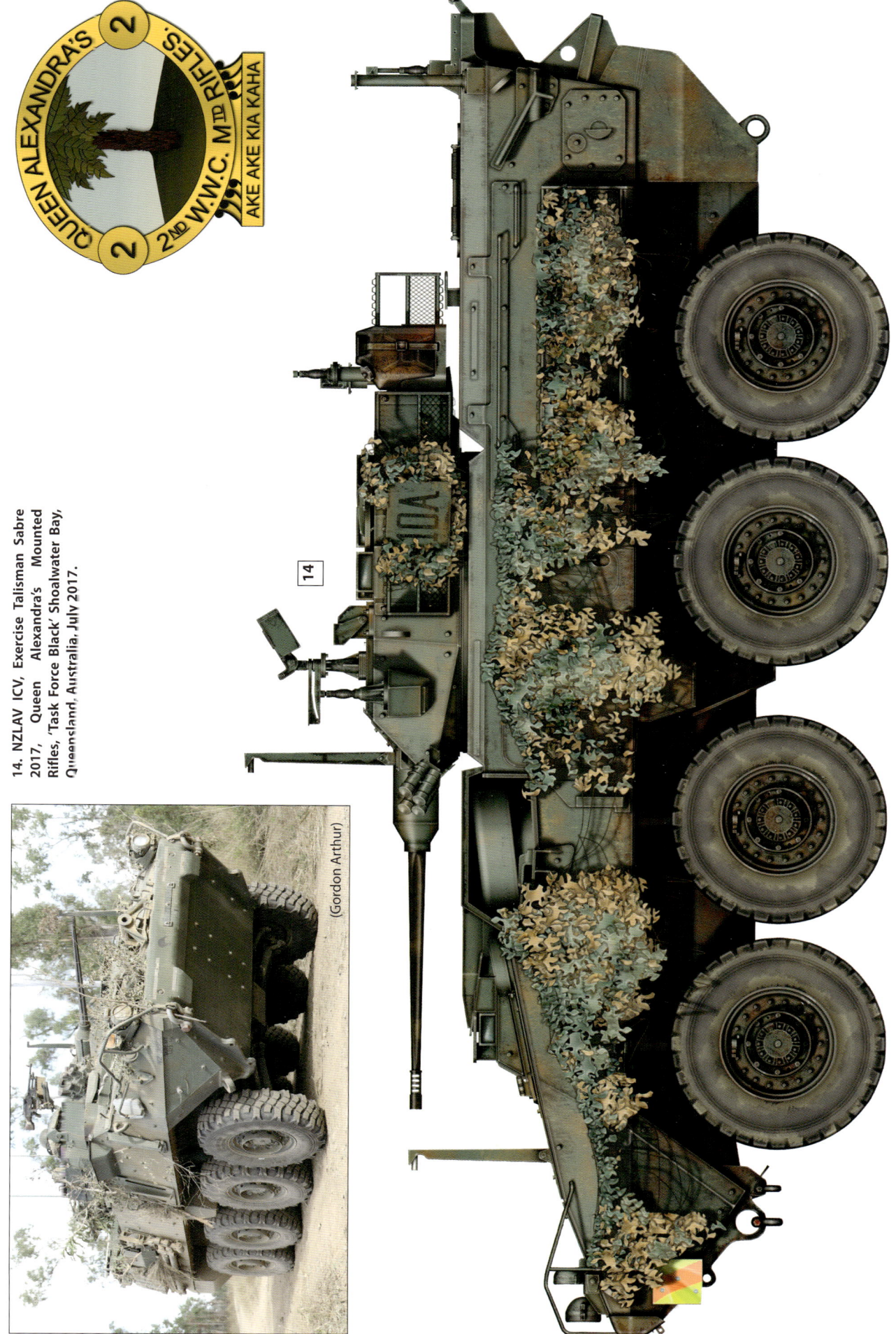

14. NZLAV ICV, Exercise Talisman Sabre 2017, Queen Alexandra's Mounted Rifles, 'Task Force Black' Shoalwater Bay, Queensland, Australia, July 2017.

(Gordon Arthur)

LAV III
3rd Battalion, Royal 22e Régiment, ISAF, Kabul, Afghanistan, 2004.
1/35 scale, Trumpeter
Carlos Blanco

Canadian LAV IIIs in a vehicle park in Kabul undergo routine maintenance at the end of their tour of duty in 2004. (Patrick Winnepenninckx)

M1132 STRYKER ESV
1st Stryker Brigade Combat Team, 25th Infantry Division, Kandahar, Afghanistan, 2011

1/35 scale, AFV Club
David Chou

(above) The assembly of AFV Club's Stryker family is straightforward if time consuming, simply because of the complexity of the real vehicle. The kit contains over 1,000 parts, some 800 of which are used in the build. There are some scratchbuilt details, for example around the M151 RWS.

(below) The M1132 with the base colour and highlights applied. The Stryker is crammed with detail and highlighting like this provides a relatively simple way of adding depth to the model even at this stage.

(above) The Surface Mine Plow (SMP) is a kit in its own right. The separately available chain and hangar set for the SMP certainly adds to the complexity of the build, but they are essential if you are after an accurate replica. Have your references to hand!

(below A view of the rear of the finished model, showing the stowage and the lane marker dispensers.

David's collection of finished AFV Club M1132s.

A M1132 ESV from 2nd Cavalry Regiment moves through a training lane at the Joint Multinational Training Area, Hohenfels, Germany, in April 2015. Note the trailer for the M58 Mine Clearing Line Charge (MICLIC). (US Army photo by Amy Weiser-Wilson)

M1134 STRYKER ATGMV

2-112 Infantry, 56th Stryker Brigade Combat Team, 28th Infantry Division, Camp Victory, Iraq, 2009.
1/35 scale AFV Club
Ramón Segarra Guerrero

(below) AFV Club's M1134 is a great kit, although complex and not without a few mistakes in the instruction manual.

(below) The Black Dog accessory set for the Stryker (ref. T35145) is very extensive. Only about half the contents of the set were used in this project.

(above) The basecoat was Vallejo Model Air NATO Green (71.093). This was then highlighted and faded using UK BSC 64 Portland Stone (71.288)

(above) An overhead view of the model prior to weathering showing the stowage. The abundance of stowage makes for a very interesting model.

Stryker ATGMVs of 2nd Cavalry Regiment take up position during a live fire exercise at the Grafenwoehr Training Area in October 2018. (US Army photo by Markus Rauchenberger)

M1128 MOBILE GUN SYSTEM
1st Squadron, 2nd Cavalry Regiment, Hohenfels Training Area, Germany, 2012.
1/35 scale, AFV Club
Sean Lynch

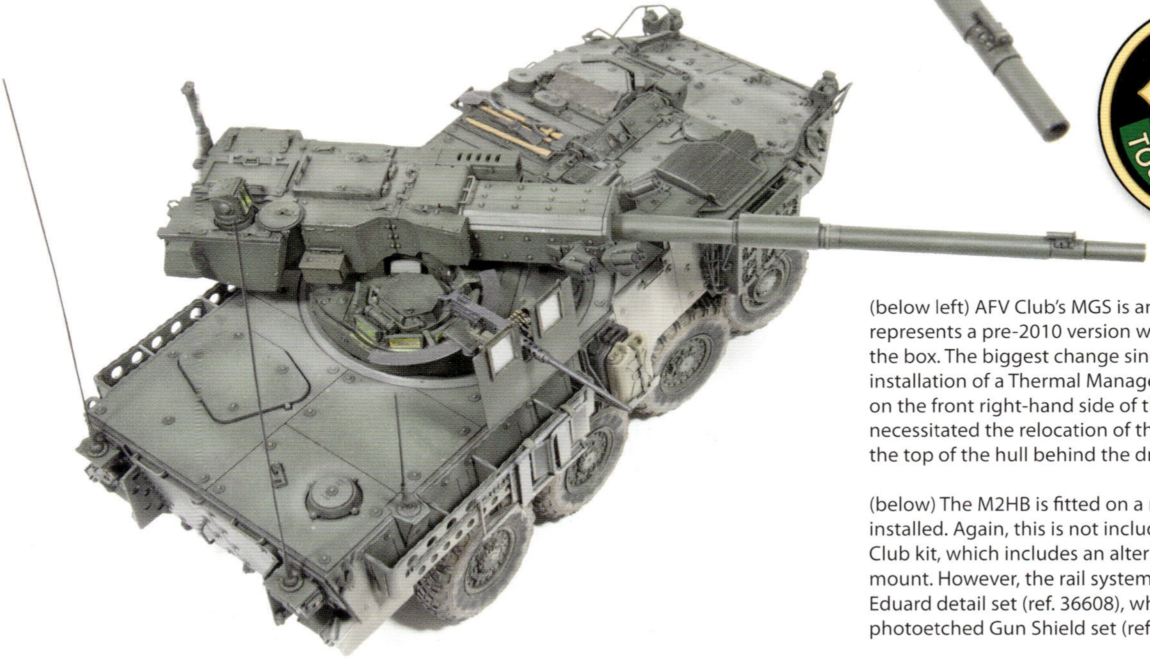

(below left) AFV Club's MGS is an excellent kit but represents a pre-2010 version when built out of the box. The biggest change since that date is the installation of a Thermal Management System (TMS) on the front right-hand side of the vehicle. This necessitated the relocation of the pioneer tool rack to the top of the hull behind the driver's hatch.

(below) The M2HB is fitted on a rail and has the shield installed. Again, this is not included with the AFV Club kit, which includes an alternative swing-arm mount. However, the rail system is included in the Eduard detail set (ref. 36608), while Voyager do a nice photoetched Gun Shield set (ref. VBS0205).

(above) The TMS is the most obvious upgrade the MGS has received since entering service. No Aftermarket version of this exists, but the one on this model was scratchbuilt from plasticard and spare photoetch, as explained in the Mar/Apr 2020 issue of AFV Modeller.

(above) The model was painted with a Tamiya acrylics. The characteristic loamy soil and dust of the Grafenwoehr Training Area has been captured perfectly with Tamiya Buff and Flat Earth.

(below) A M1128 of Apache Troop, 1-2 Cavalry, rolls across the Hohenfels Training Area in April 2012. (US Army photo by Markus Rauchenberger)

M1296 STRYKER INFANTRY CARRIER VEHICLE DRAGOON (ICVD)

Mustang Troop, 1st Squadron, 2nd Cavalry Regiment, Rose Barracks, Vilseck, Germany, 2018.

1/35 scale, AFV Club
David Chou

(above) AFV Club's Strykers have superbly detailed suspension which, with a little but of surgery, can be articulated to give a more dynamic appearance.

(above) The details on the AFV Club kit are excellent and some parts, such as the Driver's View Enhancer (in the centre to the right of the driver's hatch) are much better than that in Panda Model's kit of the same vehicle.

(below) One big omission from the Panda Model kit are the straps for the side stowage racks. They can be added, of course, using photoetch but they are included in the AFV Club, a really nice touch.

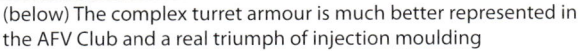

(below) The complex turret armour is much better represented in the AFV Club and a real triumph of injection moulding

M1126 STRYKER CROWS-J

Commanche Troop, 1st Squadron, 2nd Cavalry Regiment, Grafenwoehr Training Area, Germany, October 2018.

1/35 scale,
AFV Club/Blast Models,
David Grummitt

(above) To make a Stryker CROWS-J you will need the Blast Models CROWS-J RWS, as well as the AFV Club M1296 and M1126. Blast Models' also produce underbelly armour and a rear stowage shelf in case you don't have a spare M1296 in your stash.

(below) The kit wheels were replaced with those from PanzerArt, while the rear stowage shelf was from Blast Models. The aerials were from a DEF Model upgrade set for the M1A2 SEP V2 Abrams MBT.

A good rear view of a Stryker CROWS-J at the Hohenfels Joint Multinational Readiness Center in November 2019 during Exercise Dragoon Ready 20. (US Army photo by Spc. Esmeralda Cervantes)

(below) Parts of the AFV Club M151 were combined with the Blast Models' set to produce a CROWS-J.

(below) The model was sprayed with Vallejo ModelAir paints. Bronze Green 71.250 was sprayed over a Dark Green 71.012 base, which was then highlighted with Olive Yellow 71.013.

36 Stryker Interim Combat Vehicle

(left) AFV Club AF35126. (above) AFV Club 35128

The LAV III and Stryker serve in many roles in the militaries of several countries and this profusion of different versions is mirrored in the number of kits and accessories available to the model maker across different scales. The aim of this section is to survey the available kits and accessories and provide some helpful advice to modellers who wish to tackle their own LAV or Stryker project from my own experience and by synthesising the comments and experience of others.

1/35-SCALE KITS: AFV CLUB

Model manufacturers were quite slow to offer kits of the Stryker and LAV in plastic, but in 2007 both AFV Club and Trumpeter released the first of a series of kits eventually covering all the main variants of the vehicles in both Canadian and US service. AFV Club's Strykers are, without doubt, the best of the bunch, but are not without their own quirks and issues. AFV Club's first release was the M1126 Stryker 8X8 Infantry Carrier Vehicle (ref. AF35126). This is, like most AFV Club kits, a complex build with some 360 olive green plastic parts, a clear sprue for the optics, eight vinyl tyres and a sheet of photoetch, as well as poly caps and a

length of nylon twine for the winch cable. The kit is very detailed, especially the suspension (although the outer covers for the shock absorber assemblies are moulded solid, instead of perforated as they are on the real vehicle). The fit of the main hull components is excellent on all AFV Club Strykers and the majority of the build consists of placing the various external fixtures, fittings and weapons systems. The M1126 kit comes with a choice of either the .50cal M2HB or 40mm Mk. 19 Grenade launcher for the M151 RWS. This is a kit in itself and is also available separately (ref. AF35157). AFV Club's M1126 has also been reboxed by Academy (ref. 13284).

AFV Club quickly followed the M1126 with a M1130 Stryker Command Vehicle/CV TACP (Command Vehicle Tactical Air Control Party) (ref. AF35130). This is the M1126 with the additional of extra sprue containing the antennas and other extras, as well as two extra sheets of photoetch. It also replaces the original kit's solid shock absorber covers with some nicely perforated plastic ones (included in all subsequent AFV Club Stryker kits). The original boxings of the M1130 also contained a bonus resin Blue Force Tracker. These, as we have seen, are fitted to Strykers in service and it's a notable omission from most of AFV Club's other Stryker kits. AFV Club's third release, the following year, was the Stryker M1128 MGS (ref. AF35128). Naturally, this kit shares about half of the parts with the M1126, but has six new sprues, a turned metal 105mm gun barrel and a new sheet of photoetch. The kit, however, represents a prototype vehicle and MGS in action, in Iraq or Afghanistan, or in garrison have some different features so it's worth checking your references. Next up we have the M1134 Stryker ATGMV (ref. AF35134), released in 2009. Like the MGS, this kit includes a completely new upper hull, as well as new rear panel with single door. The TOW launcher is, as you might expect, a complex kit in itself, but is a wonderful replica of the real thing. AFV Club's most complex Stryker kit

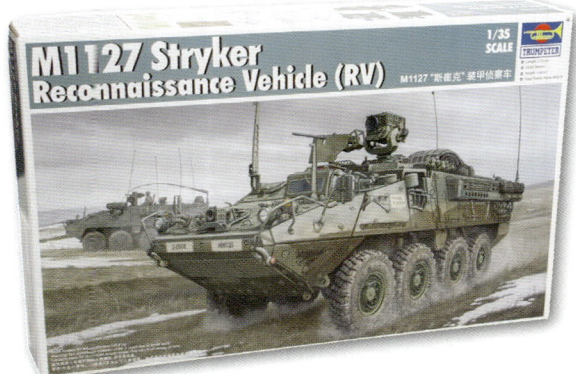

(above) Trumpeter 00395   (right) Trumpeter 01512
(below) Trumpeter 01574

(above) David Chou's build of Trumpeter's M1129 MCV-B showing the extensive interior.

was released in 2010. The Stryker M1132 Engineer Squad Vehicle SMP Surface Mine Plow (ref. AF35132) consists of almost 800 parts across fifteen sprues, as well as the usual photoetch, clear sprue and vinyl tyres. The SMP itself consists of no fewer than 296 parts and is about as complex an assembly as you'll find in plastic armour kits. That said, it is wonderfully detailed and looks superb once finished. Ideally, you'll want to add the Indicator and Chain Assembly (ref. AF35024) to the finished SMP, which is available as a separate kit. The final Stryker version to be kitted by AFV Club is the M1296 Stryker Dragoon (ref. AF35319). Again, this is superb and contains a new MCRWS turret and upper hull. The kit also contains some nice features that are useful for other 2nd Cavalry Regiment Strykers, a plastic Blue Force Tracker and additional stowage shelf above the rear door.

AFV Club's Strykers are superb kits that, apart from the complexity of the build, have little in them to criticise. Some of the assemblies are quite challenging, such as the complex headlight assembly, but just need care, patience and good references. Like all plastic kits they can be enhanced with photoetch and resin details and the vinyl tyres will not be to everyone's liking. AFV Club have also released an Upgrade Equipment for Stryker Series (ref. AF35S59) which contains UHF aerials contained on the N sprue from the M1130 kits and the N and P sprues from the M1134 (containing perforated shock absorber covers and the long tow bar often seen attached to the side of Strykers of all variants, as well as the horn mounted on the front of the hull).

TRUMPETER

Trumpeter's 1/35 Stryker kits, on the other hand, are more of a mixed bag. Trumpeter actually beat AFV Club to the market by a month in 2007 and their range of available kits is larger than their Taiwanese counterpart. Nevertheless, there are few issues that modellers should be aware of when considering a Stryker-based project. First off, Trumpeter's kits are simpler, yet still quite involved, builds compared to their AFV Club counterparts. The M1126, for example, has six grey plastic sprues, an upper and lower hull

(below) David Chou's build of Trumpeter's M1133 MEV.

and a small sheet of photoetched, as well as vinyl tyres. Some of the features, such as the M151 RWS which bears little resemblance to the real thing, are noticeably less detailed than in the corresponding AFV Club kit. The wheels and tyres are also much less realistic and have incorrect details compared to those in the AFV Club kits, while other areas (such as the headlight assemblies) are also quite poor in comparison. There is also a slight, yet noticeable, difference in the dimensions of the two kits: the AFV Cub upper hull is both narrower and longer than the Trumpeter one. On the plus side, the range of kits is much better and several include useful accessories, such as external stowage, MRE packs and a good range of decals.

Since 2007 Trumpeter have released twelve 1/35-scale kits of the Stryker/LAV III. The first was the M1126 ICV (ref. 00375), which was followed the next year by the M1127 Stryker RV (ref. 00395) and M1134 ATGMV (ref. 00399). The latter is a good example of what Trumpeter have packed into their Stryker kits: it contains eleven grey plastic sprues, a clear sprue, four figures, two sprues of personal gear, photoetch and strapping to secure the stowage, as well as a vinyl mask for the painting the driver's windshield (an option not included in the AFV Club kits), and three sheets of pre-printed drinks and MRE cartons. In 2009 Trumpeter released the M1130 Stryker CV (ref. 00397). This too can be built as the Command Vehicle or the TACP variant. The following year saw the release of the M1131 FSV (ref. 00398), as well as the LAV III (ref. 01519) and LAV III TUA (ref. 01558). These two latter kits offer a good basis for building an early variant of the Canadian LAV, but will need some work to convert them to later service versions. The same year saw the release of the M1133 MEV (ref. 01559), and two kits of the M1132 ESV, one with the LWMR-Mine Roller/SOB (ref. 01574) and one with the Surface Mine Plow (ref. 01575). In 2011 Trumpeter released a M1135 NBC RV (ref. 01560) and, finally, in 2012 the M1129 Mortar Carrier (MCV-B) (ref. 01512). This is their most comprehensive Stryker kit to date and contains a full interior, including driver's station as well as the 120mm mortar.

PANDA MODELS

In 2019 Panda Model announced the release of their M1296 Stryker Dragoon ICV (ref. 35045), bringing this to market several months ahead of AFV Club. Panda Models' kits can be a bit hit-and-miss and this is no exception. The kit comprises five sprues of dark yellow plastic, some black plastic tyres, a clear sprue and a sheet of photoetch. The sprues are, for the most part,

crisply moulded (sprues B and D particularly), but some small parts have not escaped the injection moulding process unharmed. There are some very good parts of the kit and other parts that are a real challenge. The plastic wheels and tyres, with their heavy sprue attachments, are a case in point and best replaced with some resin Aftermarket alternatives. The turret, the most identifiable part of the Stryker Dragoon, is perhaps, after the wheels, the weakest part of the kit. The kit's Bushmaster cannon really lacks detail sadly. Most of the turret assembly is straightforward, except for one crucial part: the protruding parts that cover the turret. These are prominent and distinctive on the real vehicle, but Panda would have you cut multiple tiny rods from the sprue which will test both your patience and skill. The turret also lacks the prominent anti-slip texture of the real thing. That said, it's not a bad kit by any measure; only the AFV Club version is much better.

The author's Panda Models' M1296. Note the Djiti Productions' resin Bushmaster cannon barrel.

Modelling Products

(left) Trumpeter 07255

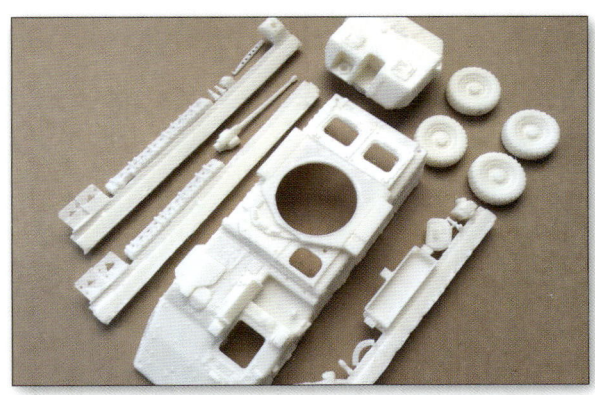

(above) The upper and lower hulls of the Academy kit (13411) are well detailed and crisply moulded.

SMALL-SCALE STRYKERS

Modellers who work in smaller scales are not well served when it comes to Strykers. There are only two plastic kits available in 1/72 scale: Trumpeter and Academy. Trumpeter released their M1126 'Stryker' ICV (ref. 07255) in 2007. The kit is nicely moulded and consists of a single grey plastic sprue and slide-moulded upper and lower hulls, as well as vinyl tyres. Most of the hatches are moulded shut, but overall it builds up into a nice replica. The following year Academy released their own M1126 Stryker (ref. 13411). This kit includes two sprues of parts besides the slide-moulded upper and lower hulls. The two-part tyres are moulded in plastic and four hatches on the upper hull can all be posed open. The detail is much sharper on this than the Trumpeter offer and it is the 'go-to' kit for modellers wanting to build the Stryker in Small Scale.

Korean-firm D Toys have released several resin conversion sets for the Academy M1126. These are the M1127 Stryker RV (ref. DT72-004), M1128 MGS (ref. DT72-005) and M1134 Stryker ATGM (ref. DT72-006). These are expensive, but very well made. Cromwell Models also produce resin conversions for the Academy kit. These are the Stryker Dragoon (ref. CA017), which includes a complete new upper hull, turret, rear stowage shelf, and four wide-pattern tyres, and the Stryker MGS (ref. CA018). This includes upper hull, tyres, turret and 105mm gun, a new rear panel, tyres and other details. German-firm Modell Trans Modellbau also offers reasonably priced and very good conversions for the M1127 RV/M1131 FSV (ref. MT72150), M1128 MGS (ref. MT72148), M1129 MC (ref. MT72151), M1130 CV (ref. MT72152) and M1133 MEV (ref. MT72147).

Several firms offer resin replacement wheels for the Academy kit: Armory (ref. AC738), D-Toys (ref. DT72-001), DEF.Model (ref. DW72003), MR Modelbau (ref. MR-72141) and Modell Trans Modellbau (ref. MT72123). Both E.T. Model (ref. E72-026) and Eduard (ref. 22127) offer photoetch detail sets for the Academy kit, while Eduard also produce a set of slat armour (ref. 22129). Legend Productions (ref. LF72080) and Black Dog (ref. T72002) also offer dedicated resin stowage sets for 1/72-scale Strykers.

(above) Cromwell Models CA017

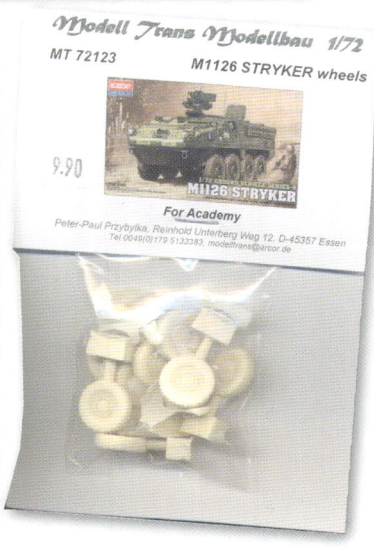

(above right) D Toys DT72-001
(right) DEF Model DW72003
(far right) Modell Trans Modellbau MT72123

ACCESSORIES: WHEELS

The eight Michelin XML 1200R20 tyres are such a prominent feature on the Stryker IAV that it is no surprise that many modellers look to resin replacements for the vinyl tyres contained in the AFV Club and Trumpeter kits. Several companies have released replacement wheel sets, consisting of Hutchinson steel rim and Michelin tyre, over the years, but several of these are now out of production and difficult to find. Armorscale (ref. R35-050), E.T. Model (ref. ER35-010 for Trumpeter, ER35-011 for AFV Club and ER35-012 for the LAV III), Pro Art Models (ref. PAU-35027), Real Model (RM35143 for AFV Club, RM35151 for Trumpeter and RM35144 for LAV III), Tiger Productions (ref. 35003), TMD (ref. 35-73901), Voyager Models (ref. PEA094 and PEA423 for the M1296 Stryker Dragoon) all produce or have produced resin wheel sets, but the most widely available and best, in terms of both value and quality, are the sets by Panzer Art (ref. RE35-123) and DEF Model. DEF Model are the gold standard for resin replacement wheels, but are expensive. They have produced three sets for the Stryker/LAV: ref. DW35010A, superseded in 2010 by DW35010A, and ref. DW35114 for the M1296. Worth mentioning too is the excellent Djiti's Production set for Panda Model's M1296 (ref. 35091). This includes not only eight wheels but also a superb 30mm XM183 cannon.

PHOTOETCH

Modellers who have worked with photoetch either love it or hate it. On the one hand it allows oversized details in plastic kits to be replaced or the addition of those missing altogether. On the downside, it is necessarily two-dimensional and can be fiendishly difficult to assemble. The Stryker and LAV family are complex and 'busy' vehicles and the various photoetched detail sets are accordingly complicated. The largest producer of detail sets for the Stryker family, many of which are now out of production but still available from various online retailers, is the Czech-firm Eduard. They produce both specific sets for the various kits, both AFV Club and Trumpeter, but also generic sets such as Stryker Stowage Belts (ref. 36091), Slat Armour (ref. 35995 for Trumpeter and ref. 36002 for AFV Club), and Additional Armour (ref. 36015 for Trumpeter and ref. 36011 for AFV Club). Many of these vehicle specific sets have been combined with the generic sets in Eduard's 'Big Ed' sets. The Eduard sets are very useful and contain some items that are not included in the plastic kits, but they are challenging and only for the experienced modeller.

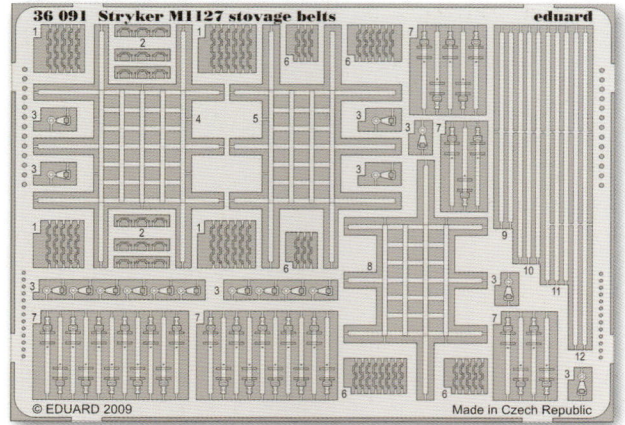

(left) DEF.Model DW35114
(above) Eduard 36091
(below) Real Model Ltd OIF Accessories Set

Modelling Products   41

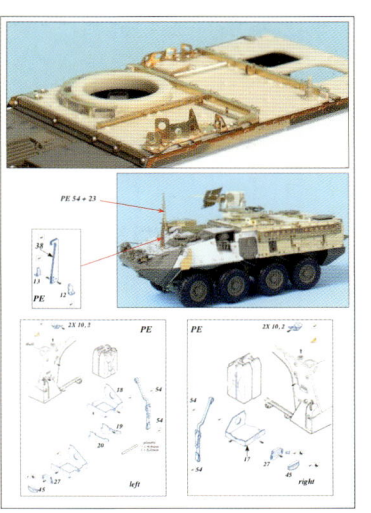

(left) Real Model RM 35146
(above) Black Dog T35053
(below) Blast Model BL35114K

More recently Chinese-firms E.T. Model and Voyager have taken a slightly more sophisticated approach in producing photoetched detail sets which contain different thicknesses of photoetch, as well as resin, brass and plastic rod. E.T. Model's range is limited to the LAV III (ref. E35-050), M1126/M1130 (ref. E35-222) and M1132 (E35-223), as well as a set of LAV III Smoke Dischargers (EA35-056). Voyager's range is much more extensive, including both vehicle-specific and generic sets. The latter include a beautiful M151 RWS (ref. PEA093) and M2HB and Gun Shield for the M1128 (ref. VBS0205), as well as Slat Armour for the M1126 (ref. PE35200). Voyager also produce a set of pre-coloured headlight and tail light lenses for the AFV Club (ref. BR35126). These lenses really lift the appearance of the kit, as do those available from SKP Model (ref. SKP116).

CONVERSIONS AND DETAIL SETS

Several firms offer resin detail and conversion sets to update or convert the AFV Club and Trumpeter kits. Czech-firm Real Model led the way on this with upgrade sets for many of the Trumpeter kits and an extensive range of conversions for the LAV III based on the AFV Club Stryker, as well as upgrades of the Trumpeter LAV III These included the LAV III LORIT (ref. RMA35215), LRAS 3 (ref. RMA35235) and a LAV III Command Post (ref. RM35132). The Real Model sets combine photoetch made by Eduard (some of it unique to the Real Model kits) with their own resin. Another Czech-firm Black Dog also do a nice range of resin accessories and conversions for the Strykers. Their accessories sets are a nice alternative to photoetch and are combined with the WIN-T (Warfighter Information Network-Tactical) conversion for the M1130 (refs. T35145, T35146, T35147, T35148, T35149). They also produce an M1126 interior (ref. T35001) and accessories sets for both the basic LAV III (ref. T35033) and LAV III LORIT (ref. T35053).

French-firm Blast Model also produce a good range of resin accessories for the Stryker. These include a couple of nice sets of generic resin stowage (refs. BL35114K and 115K) and the new-style rear stowage shelves seen on 2nd Cavalry Regiment Strykers (refs. BL35370K, BL3571K). They also do a very useful range of sets allowing you to convert the AFV Club kits to the latest specification. These include Blue Force Trackers (ref. BL3558K), the DEK Driver Enhancement Kit (ref. BL35369K), the M151A2 RWS (ref. BL35146K) and CROWS-J RWS (ref. BL35357K). Blast Models are reasonably priced and good quality. Korean-firm Legend Productions' also produce a very large stowage set for Strykers (ref. LF1153), as well as engine sets for both the AFV Club (ref. LF1217) and Trumpeter kits (ref. LF1220).

(left) Blast Model BL35358K

Legend Productions LF1153

(above) An MGS from 5th SBCT, 2nd Infantry Division, fires its main gun in Hutal, Afghanistan in January 2010. Note the slat armour, designed to protect against Rocket-Propelled Grenades, fitted to many Strykers in Afghanistan and Iraq. (US Army photo by Staff Sgt. Dayton Mitchell)

(below) An MGS from Ghost Troop 3-2 Cavalry during Exercise Iron Panzer in August 2011. Note the shield for the commander's M2HB and the air-conditioning unit fitted to the sides of all M1128s from 2010. (US Army photo by Gertrud Zach)

M1128 STRYKER MOBILE GUN SYSTEM (MGS)

The MGS mounts the same M68A1E4 rifled 105mm gun as the M1 Abrams Main Battle Tank. It entered service with 4-23 Infantry, 4th SBCT, in 2006, three years after the other Stryker variants. In all 142 MGSs were procured by the US Army. Plans for it to be adopted by the Canadian Army as a replacement for its tank fleet were abandoned in 2007. The MGS was initially intended to provide direct fire support to dismounted infantry during the assault. Its primary function was envisaged to be the destruction of enemy bunkers, machine gun nests and sniper positions, as well as creating breaches in urban situations. Its anti-armour capabilities were initially seen as limited and secondary to its infantry support role. It carries eighteen rounds of ammunition for the 105mm gun and can fire high-explosive, high-explosive anti-tank (HEAT), high-explosive plastic tracer (HEP-T), armour piercing discarding sabot (APDS) and canister. The 105mm gun has an effective range of 2km and rate of fire of eight rounds per minute. The gun has an autoloader, which accommodates eight rounds with ten rounds stored elsewhere in the vehicle. The main gun is controlled by a Fire Control System and is fully stabilised, allowing the MGS to fire on the move. It is also armed with an M2HB on the commander's cupola and coaxially mounted M240C 7.62mm machine gun. Initially 27 MGSs served in each SBCT, a platoon of three vehicles for each infantry company (nine MGSs per Stryker infantry battalion).

The development of the MGS, like other Stryker variants, was not without its problems. The first prototypes, at 52,000lbs, were too heavy and too tall to be carried in the C-130 aircraft, while the vehicle was prone to rolling over if the main gun was fired when rotated through 90 degrees. Early deployments in Iraq identified other issues that made the MGS a problematic, although very valuable, component of the SBCT. First, the 105mm

gun was very loud: hearing protection was necessary for all soldiers up to 25m behind the muzzle to prevent perforated eardrums, while vehicles within that range had to be 'buttoned up' to prevent harm to crew members. The back blast extended to 100m and anyone behind the gun was in danger of being hit by ejected cartridges. At a tactical level, the MGS platoon was rarely able to operate as such; with only a crew of three (commander, gunner and driver), it proved impossible to provide sufficient vehicle security within urban areas and the MGS was usually fielded singly in support of infantry platoons. Moreover, in the field no SBCT had a full complement of MGSs until 2009, with the M1134 ATGMV Stryker and ICV filling the gaps in those platoons that could not muster three MGSs.

In the years following the MGSs' combat debut in Iraq its role in the SBCT has changed. There were plans to reduce the number of MGSs in each brigade from 27 to just ten, but in 2017 a new organisation was introduced of twelve M1128 per SBCT organised either in three four-vehicle platoons or as three platoons with two M1128 and one M1134 and three with one M1128 and two M1134. This allows greater flexibility in the MGSs tactical employment and reflects the shift toward planning for deployment against a peer or near-peer adversary.

(above) An MGS from Fox Troop, 2-2 Cavalry leads an ICV onto the ranges at Grafenwoehr Training Area in March 2012. (US Army photo by Staff Sgt. Jose Ibarra)

(below) A MGS of 3rd Cavalry Regiment, 'the Brave Rifles', at Fort Hood, TX, in 2015. (US Army photo by Major Vance Trenkel)

(below) An MC-B of Mustang Troop, 1-2 Cavalry fires its M252 mortar in Germany in January 2017. (US Army photo by Gertrud Zach)

(bottom) Another MC-B, this time of Head Hunter Troop, 2-2 Cavalry prepares to leave a range at the Bemovo Piskie Training Area, Poland, in June 2017 during Exercise Saber Strike 17. (US Army photo by Spc. Samuel Brooks)

M1129 STRYKER MORTAR CARRIER VERSION

The mortar carrier serves with both the mortar platoons of the infantry battalions within the SBCT and as part of the mortar section of the combat companies. The latter also carries a 60mm M224 Lightweight Mortar, while the former has an additional 81mm M252 Medium Extended Range mortar. There are also six assigned to the RSTA battalion, although these carry only the 120mm mortar. Each SBCT thus has 36 mortar carriers. The original M1129 MC-A was simply a M1126 designed to carry a M121 120mm mortar in the rear of the ICV. The crew of five (commander, driver, gunner, assistant gunner and loader) had, however, to dismount to carry out fire missions. From 2005 the MC-B entered service carrying the Recoiling Mortar System (RMS) 6L 120mm mortar on a rotating turntable in the squad compartment (although the mortar can only fire to the rear of the vehicle). The MC-B is not fitted with the M151 RWS and instead has a rail-mounted M240B 7.62mm machine gun. The mortar can fire a range of munitions, including smoke, white phosphorous and high explosive. It has an effective range of 6.7km and can sustain a rate of fire of four rounds per minute. The MC-B is equipped with the M95 Mortar Fire Control System, meaning it is fully integrated into the brigade's digital network and capable of receiving fire missions on the move.

M1130 STRYKER COMMAND VEHICLE (CV)

Effective command and control, linking troops on the ground to the rest of the combined arms force, are an essential part of today's digital battlefield. In the SBCT this role is filled by the M1130 CV. The CV is essentially an ICV with the infantry squad compartment filled with various command and control (C4) equipment. The exact configuration of the CV depends on its role within the SBCT: the Brigade Commander Command Vehicle, Battalion Commander Command Vehicle, S3 TAC (attached to the Brigade Signal Company), RSTA Squadron Commander Vehicle, Battalion S3 (attached to each Infantry Battalion), and Tactical Air Control/USAF. In all 25 CVs are issued to each SBCT. The range of communications equipment that can be housed within the CV is impressive: it includes a Demand Assigned Multiple Access line-of-sight and tactical satellite communications terminal and a Near-Term Digital Radio (NTDR) system that can network up to 400 individual radios across a 20x30km area.

Externally the CV can only be distinguished from the ICV by the various antenna arrays. It is armed with the M151 RWS and also carries a Javelin ATGM Command Launch Unit, as well as a single M136 AT-4 light anti-tank weapon. It has a crew of five (commander, driver, staff officer and two workstation operators), although an additional workstation seat has subsequently been added. The CV has been upgraded with the double v-hull design.

(above) M1130 CVs of 1st Battalion, 23rd Infantry Regiment, 3rd SBCT, 2nd Infantry Division in Tal Afar, Iraq, in September 2004. Note the slat armour designed to counter RPGs and additional antennae of the Command Vehicle. (NARA: Sgt. Jeremiah Johnson)

(below) A CV of 2-2 Cavalry during Exercise Allied Spirit IV held at the Joint Multinational Readiness Center at Hohenfels, Germany, in January 2016. (US Army photo by Sgt. William A. Tanner)

M1131 STRYKER FIRE SUPPORT VEHICLE (FSV)

The M1131 is issued to the SBCT's Anti-Armor Company, RSTA Squadron, and Infantry Battalions. The crew of four (commander/sensor operator, driver, mission specialist and an additional crew member) direct, coordinate and execute fire support operations from the SBCT's mortars and the Fires Battalion's M577 155mm howitzers (which replaced the M198 howitzer in US service in 2005). The FSV was initially equipped with the AN/TAS-4 and AN/TVQ-2 Ground Vehicular Laser Locator Designator (G/VLLD) mounted on the power-assisted commander's cupola, but this has been replaced by the more sophisticated Fire Support Sensor System (FS3) Mission Equipment Package (MEP) with Laser Designator Module. This allows the FSV to order indirect fire support with conventional, GPS and laser-guided munitions. Externally, the FSV is indistinguishable from the RV except for the additional antenna arrays. Twelve FSV serve with each SBCT.

(above) An M1127 FSV of 4th SBCT, 2nd Infantry Division on the Iraq/Kuwait border in August 2010. Only the additional antennas differentiate the FSV externally from the RV. (US Army photo by Sgt. Kimberley Johnson)

(below) FSV of Nemesis Troop, 2-2 Cavalry on the move at the Grafenwoehr Training Area during Exercise Saber Strike in October 2012. (US Army photo by Spc. Jordan Fuller)

M1132 STRYKER ENGINEER SQUAD VEHICLE (ESV)

The M1132 ESV is the mainstay of the Brigade Engineering Battalion within the SBCT. Each Combat Engineer Company has twelve such vehicles. The ESV is basically an ICV with the Jettison Fitting Kit (JFK) mounted on the front of the vehicle. This can be fitted with one of the following specialist engineering attachments: the Pearson Engineering Surface Mine Plow (SMP), Lightweight Mine Roller (LWMR), Angled Mine Plow (AMP) or Straight Obstacle Blade (SOB). They are controlled by the Mission Equipment Control Unit in the driver's compartment and can be hydraulically raised and lowered. Each of the attachments has a specific use: the SMP and AMP are used to detonate surface mines, while the LWMR is used to detonate buried mines, clearing a path 3.14m wide (with an uncleared central section of 1.19m). The SMP can clear a path up to 4.5m wide, while the AMP can clear a path 4.2m wide. These attachments can also be fitted with the Magnetic Signature Duplicator which detonates magnetically fused mines ahead of the rollers or blades. The SOB is used to clear obstacles and move earth and can also be used in the preparation of defensive positions. To the rear of the vehicle are two Lane Marking Equipment dispenser units which shoots metre-long poles into the ground to demarcate the cleared path. One ESV in each company also carries a TALON military robot for explosive ordnance disposal and for disarming Improvised Explosive Devices (IEDs). The ESV can also tow the trailer-borne version of the M58 Mine Clearing Line Charge (MICLIC), a rocket-assisted explosive line charge designed to clear a 100m-long path through minefields 8m in width.

The ESV is protected by the M151 RWS and carries a full nine-man squad (albeit of combat engineers). The ESV also carries a Javelin Command Unit, 7.62mm M240B machine gun and two M136 AT-4 light anti-tank weapons, as well as the squad's personal weapons. Above the rear exit ramp

(top) An ESV fitted with the LWMR and slat armour in Iraq. Note the anti-IED antenna just visible behind the RWS and the improvised sun shelter. (US Army photo by Spc. Joshua Edwards)

(left) A good view of the SMP fitted to a ESV of 2nd Cavalry Regiment during Exercise Sabre Junction in October 2012. (US Army photo by Gertrud Zach)

of the vehicle is an extended stowage basket which carries much of the other engineering equipment, including demolition equipment and the Squad Pioneer Tool Kit. The ESV has been upgraded with the double-v hull.

(right) Another 2nd Cavalry Regiment ESV, this time fitted with the LWMR. (US Army photo by Gertrud Zach)

(centre) An ESV of 2nd Cavalry Regiment equipped with a SOB leaves the Czech Republic en route for Poland during Exercise Dragoon Ride in May 2016. (US Army photo by Spc. Sandy Barrientos)

(bottom) A good rear view of ESVs of 2nd Cavalry Regiment, showing the Lane Marking Dispensers, as they prepare to move on the Bemowo Piskie Training Area, Poland in June 2017 during Exercise Saber Strike 17. (US Army photo by Samuel W. Brooks)

M1133 STRYKER MEDICAL EVACUATION VEHICLE (MEV)

The MEV is unarmed and four are issued to the headquarters companies of the RSTA and Infantry battalions of the SBCT. It allows the SCBT to safely evacuate and treat wounded soldiers from the frontline. In addition to the crew of a commander, driver and paramedic, the squad compartment is modified to accommodate either six seated or four stretchered patients or a mixture of three seated and two on stretchers. The vehicle is unarmed, except for the M66 grenade system which fires smoke grenades. The roof of the troop compartment of the MEV is raised by a height of 250mm. The vehicle carries all the necessary medical equipment to treat and stabilise injured soldiers at the front. It is also fitted with the Medical Communications for Combat Casualty Care (MC4) digital system that can provide medical records for patients, as well as medical situational awareness for commanders. Each SBCT has sixteen such vehicles.

(top) The raised roof of the MEV compared to the standard ICV is evident from this image of an M1133 returning from Exercise Hanuman Guardian in Thailand in July 2016. (US Navy photo by Grady Fontana)

An MEV of Mustang Troop, 1-2 Cavalry on the Bemowo Piskie Training Area during Exercise Saber Strike in June 2018. (US Army National Guard photo by 1st Lt. Erica Mitchell)

M1134 STRYKER ANTI-TANK GUIDED MISSILE VEHICLE (ATGMV)

Originally the ATGMV was the principal weapon of the SBCT's Anti-Armor Company although, as we have seen, it now shares that role alongside the MGS. The ATGMV has the Elevated TOW System (ETS) fitted instead of the RWS. The BGM-71 TOW (Tube-Launched, Optically Tracked, Wire-Guided) missile is one of the longest-serving weapons systems in the US Army, first entering service in 1970. The ETS can be fitted with either the BGM-71F TOW-2B anti-armour missile or the BGM-71H TOW-2A 'bunker buster'. The former is capable of penetrating the armour of all known AFV and has a maximum effective range of 4.5km. It has a 'top-down' attack capability, allowing it to penetrate the thinner top armour of enemy AFVs. Each TOW-2B has

(above) A 2nd Cavalry Regiment ATGMV at the Joint Multinational Readiness Center during Exercise Saber Strike in October 2012. (US Army photo by Sgt. Ian Schell)

(right) Another 2nd Cavalry Regiment ATGMV photographed later that month during Exercise Saber Junction. Note the M240B mounted on the commander's cupola. (US Army photo by Markus Rauchenberger)

two warheads with Explosively Formed Penetrators (EFP), one which detonates directly forwards (or downwards) and one at a slight angle, designed to defeat Explosive Reactive Armour (ERA) tiles. The 'Bunker Buster' TOW carries a high-explosive charge which can penetrate up to 200mm of reinforced concrete at effective ranges of up to 3km. Ten spare missiles, along with the two loaded in the launcher, can be carried in each ATGMV and the vehicle is served by a crew of four (commander, loader, gunner and driver). The ETS can be rotated through 360 degrees and incorporates a Modified Improved Target Acquisition System (MITAS) which controls target detection, acquisition and fire control. It is equipped with both day and night vision cameras. The Fire Control System tracks the missile to its target by allowing the gunner to lock onto the enemy AFV's thermal signature.

The ATGMV must, however, must remain stationary when engaging targets (and the TOW missile must be tracked for 23 seconds to a target 4.5km away). Secondary armament is the form of a M240B machine gun mounted on the commander's cupola, although this must be folded away when the ETS is in use leaving the vehicle vulnerable to enemy infantry. The ETS itself can be lowered by 550mm to facilitate air travel.

Before the full-scale deployment of the MGS, the ATGMV also served in the infantry battalions with as many as 36 vehicles in each SBCT. Under the 2012 Table of Organisation this was substantially reduced and the Anti-Armor Company had nine ATGMVs in three platoons. In 2017 the ATGMVs were again reorganised into the SBCT Weapons Troop of three three-vehicle platoons or six platoons of mixed ATGMVs and MGSs.

(left) A good image of a 2nd Cavalry ATGMV showing how the ETS must be elevated to 40 degrees to allow manual reloading of the TOW missiles. Tactically this is one of the limitations of the system. (US Army photo by 1st Lt. Ellen Brabo)

(below) A TOW missile is fired from an ATGMV of 1st SBCT, 4th Infantry Division, at Fort Carson at Fort Carson, CO, in March 2020. (US Army photo by Sgt. Micah Merrill)

(right) An NBCRV of the 690th Chemical Company, Alabama Army National Guard, at the Yakima Training Center, WA, in April 2019. Mounted on the rear of the vehicle is the automated sample collector for the Chemical Biological Mass Spectrometer Block II (CMBS II). (US Army photo by Pfc. Valentina V. Montano)

(below) An NBCRV of 1st SBCT, 1st Armored Division checks radiation levels at the White Sands Missile Range in February 2019. This is a useful exercise that allows the NBCRV to calibrate its readings against those generated by a computer model. Next to the RWS, protected by a canvas cover, is the Joint Services Lightweight Standoff Chemical Agent Detector. (US Army photo by John Hamilton)

M1135 STRYKER NUCLEAR BIOLOGICAL CHEMICAL RECONNAISSANCE VEHICLE (NBCRV)

First fielded in 2006 and approved for full-rate production in July 2007, the NBCRV provides the SBCT with the means to identify NBC threats. It contains various chemical and biological agent detector systems, as well as the ability to collect soil samples and lay tape and flags to designate contaminated areas while in a sealed NBC proof state.

The NBCRV has a crew of four (commander, driver, surveyor and assistant surveyor). The contamination data collected by the crew can be fed back to the rest of the SBCT and to higher formations using the NBC Sensor Processing Group (NBCSPG) by the commander and surveyor. The NBCRV is fitted with a complete NBC protection system, supplying the vehicle with clean, cooled air allowing it to operate in any environment. Three serve in the CRBN Platoon of the RSTA squadron in each SBCT, as well as with the ABCT.

M1296 STRYKER INFANTRY CARRIER VEHICLE DRAGOON (ICVD)

As we have seen, one of the complaints directed towards the Stryker IAV is its lack of firepower with which to engage a peer or near-peer adversary. This only became more acute in the aftermath of the Russian intervention in Ukraine in 2014. Consequently on 30 March 2015 Colonel John V. Meyer of the 2nd Cavalry Regiment submitted an Operational Needs Statement to the House and Senate Armed Services Committee requesting that 81 of his ICVs be fitted with 30mm cannon. Originally there had been moves for a fleet-wide upgrade of the Stryker, but Colonel Meyer's statement prioritised the change for the only SBCT based in Germany. In September 2015 the US Department of Defense signed an $8 million contract with General Dynamics Land Systems, expanded to $75 million in January 2016 after the

(above) An ICVD of 1-2 Cavalry takes part in a live fire exercise during Operation Kriegsadler at the Baumholder Military Training Area, Germany, in February 2019. (US Army photo by Erich Backes)

(below) A good view of the 30mm XM183 cannon and the sophisticated optics of the MCRWS. (US Army photo by Gertrud Zach)

(above) An ICVD of Bull Troop, 1-2 Cavalry. This one was photographed during Exercise Saber Guardian in Romania in June 2019. (US Army photo by Spc. Joseph Knoch)

(above) The principal task of the ICVD remains to transport the infantry to engage the enemy or hold ground. This heavily laden ICVD of Commanche Troop, 1-2 Cavalry was photographed at the Joint Multinational Readiness Center in November 2019. (US Army photo by Ethan Valetski)

first prototype had been delivered. In the new fiscal year Congress approved $411 million to upgrade 93 Stryker ICVs to the new configuration. The main challenge was to incorporate a 30mm cannon and retain room for a full nine-man infantry squad. In December 2015 GDLS chose the Kongsberg Protector 30mm MCT Medium Calibre Remote Weapon Station (MCRWS) to fulfil this requirement. The MCRWS is armed with a 30mm XM183 automatic cannon, a version of the tried-and-tested Mk 44 Bushmaster, and carries 156 rounds of linkless, ready-to-fire rounds of either fin-stabilised discarding sabot or high explosive incendiary, as well as training rounds. The turret can be reloaded from the inside and the 30mm cannon is effective at ranges up to 3km. The MCRWS is equipped with a CCD video day sight, thermal imager and eye-safe laser rangefinder. The increased weight of the MCRWS has also necessitated a redesign of the roof of the ICV and wider tyres

Currently half of the ICVs in 2nd Cavalry Regiment are equipped with the MCRWS and it has proved a great success. In March 2019 the US Army announced that a new A1 Medium Calibre Weapon System will be fitted to sufficient double-v hulled Strykers (the current ICVD is based on the flat-bottomed Stryker) to equip three of the six US-based SBCTs with the first receiving its vehicles in 2022.

(above) The IM-SHORAD prototype, armed with a 30mm cannon as well as Stinger and Hellfire missiles.

OTHER STRYKER VARIANTS

One of the great advantages of the LAV III design was its versatility and the US Army has demonstrated that in its development of the Stryker IAV to fulfil the full range of combat tasks, from Command and Control to engaging enemy armour. Other platforms are currently under development, including most importantly an Interim Manuever Short-Range Air Defense (IM-SHORAD) version of the Stryker. This will include a RWS armed with both Longbow Hellfire and Stinger missiles, as well as a 30mm XM914 cannon and M240B 7.62mm machine gun. It can be used against rotary and fixed-wing aircraft, as well as Unmanned Aerial Vehicles (UAVs). A Counter-Unmanned Aerial Systems (C-UAS) capability is also being developed to detect and defeat small and medium-sized drones. The C-UAS Mobile Integrated Capability (CMIC) mounts an electronic-scanning tracking radar to detect targets and a 30mm cannon to destroy them. The aim is to replace the anti-aircraft capability in the SBCT currently provided by Humvee-mounted AN/TWQ-1 Avenger and man-portable Stinger missiles with a twelve-strong SHORAD company. Prototypes of these vehicles have been delivered to the US Army for testing and they are expected to be fielded with Stryker units by 2022.

One of the more interesting Stryker variants is the Stryker Mobile Expeditionary High Energy Laser (MEHEL). The US Army plans to use to lasers to defeat enemy UAVs. The Stryker MEHEL will also include an array of electronic warfare (EW) jamming equipment designed to scramble drone systems. The system was first successfully tested in 2016, when a 2kw laser mounted on a Stryker downed 21 small quadcopter drones. In April 2018 the US Army deployed a 5kw MEHEL 2.0 Stryker during Exercise Combined Resolve in support of the 2nd Armored Brigade Combat Team to further test the capabilities of laser-based weapons systems.

A Mobile Expeditionary High Energy Laser (MEHEL) 2.0 Stryker during a tactical road march, part of Exercise Combined Resolve X at the Grafenwoehr Training Area in April 2018. (US Army photo by Pfc. Maximilian Huth)

(right) A LAV III in Kabul in 2004. Note the Bushmaster cannon has been temporarily removed and rests on the front of the hull. (Patrick Winnepenninckx)

LAV AROUND THE WORLD: CANADA

Next to the United States of America, Canada is the largest user of the LAV III/Stryker family of vehicles. In August 1995, after several years of debate over how the Canadian Army would replace its M113 Armoured Personnel Carrier and Armoured Vehicle General Purpose (AVGP) fleet, the Canadian government awarded a contract to General Motors Diesel Division (subsequently GDLS) to produce the LAV III. Alongside the Coyote LAV II Reconnaissance Vehicle (based on the LAV-25), the LAV III provides the Canadian armed forces with the full spectrum of combat roles, from transporting the infantry safely into combat, to C2 and combat engineering. The basic LAV III has a crew of three (a driver, commander and gunner) and can carry seven infantry. It is armed with an M242 25mm Bushmaster chain gun and coaxial 7.62mm machine gun. Another 5.56mm or 7.62mm machine gun can be fitted on the commander's cupola.

(below) A LAV 6.0 of the Royal Canadian Regiment trains alongside Italian troops as part of the Latvian-based Multinational Battalion Battlegroup, part of NATO's Enhanced Forward Presence, in 2018. (USMC Photo by Lance Cpl. Angel D. Travis)

The turret is equipped with a Thermal Imaging System and Tactical Navigation System, as well as an LCD monitor to show the passengers real-time battlefield images. Introduced into service on 1999, the LAV III had its baptism of fire in Afghanistan in 2003 and remained a core part of the Canadian contingent of ISAF until 2011. Initially, the infantry battalions fielded unmodified LAV IIIs, but, as we shall see, the danger of IEDs forced an urgent need to modify the LAV III. This resulted in the LAV III LORIT (LAV Operational Integration Task) from 2009. As well as additional under-belly armour, some LAV IIIs previously fitted with the anti-tank TUA (TOW Under Armour) turret were fitted with the Nanuk Remote Weapon Station (RWS) and referred to as LAV Infantry Section Carrier (LAV ISC). By March 2009 33 LAV ISCs had been issued to the Canadian ISAF contingent.

In the aftermath of the Afghanistan deployment, the Canadian LAV III fleet required a comprehensive overhaul. Thirteen had been lost to enemy action, while another 159 had been damaged. The Canadians considered replacing their entire LAV III fleet but in October 2011 announced that GDLS had been awarded a $1.1 billion contract to upgrade 550 LAV III vehicles. The upgrades relate principally to improved protection against IEDs and mines. This is in the form of the same Double-V hull fitted to upgraded Strykers. The upgrade programme also include a new 450hp engine, steering and braking systems, as well as improved fire control, thermal imaging and digital data systems. The first upgraded LAV IIIs, known as the LAV 6.0, were delivered in February 2013 and they saw their first operational deployment as part of NATO's Enhanced Forward Presence

(above) A LAV 6.0 takes part in Exercise Common Ground II at the 5th Canadian Division Support Base (5 CDSB) Gagetown in New Brunswick in November 2019. (Corporal Cayer, 2 R22eR)

(below) A LAV 6.0 CSV MR vehicle during Exercise Common Ground in November 2019. (Corporal Cayer, 2 R22eR)

(above) The vehicle of the CO of Queen Alexandra's Mounted Rifles (note the 'QAMR' number plate) during an exercise in Canterbury, New Zealand in 2011. (Gordon Arthur)

(below) A NZLAV Recovery Vehicle (RV) during Exercise Kiwi Koru. The RV has a main winch and dozer blade (with vice attached). (Gordon Arthur)

in Latvia in 2017. In 2016 GDLS developed the LAV 6.0 Combat Support Vehicle (CSV). The LAV 6.0 CSV is equipped with an M153 CROWS II RWS and can be adapted to the command, ambulance and maintenance and recovery (MR) roles. The LAV 6.0 will also eventually replace the Coyote Reconnaissance Vehicle and the remaining LAV II Bisons and M113s in Canadian service. The LAV upgrade programme was due to be completed in December 2019 at a total cost of $1.8 billion.

NEW ZEALAND

In January 2001 the New Zealand signed a contract for the purchase of 105 LAV III (102 infantry carriers and three recovery vehicles) from GDLS for a total of $653 million. The first vehicles entered service two years later. The NZLAV has basically the same specifications as the LAV III in Canadian service: it is armed with the M242 Bushmaster cannon, two 7.62mm machine guns and turret-mounted smoke dischargers and has the same passenger capacity as the LAV III. It can also be fitted with under-belly armour as a protection against IEDs, as well as add-on appliqué armour.

From the beginning the purchase of the LAV III was controversial with questions around the suitability of the LAV III platform for the terrain and the role the New Zealand army was likely to be deployed in. In the late 1990s the Army had pressed for the purchase of the LAV to replace the ageing M113 fleet and won the

l(left) An Infantry Carrier Vehicle (ICV) participating in Exercise Talisman Sabre 2015. This was first time that NZLAVs joined this large multinational exercise in Australia. (Gordon Arthur)

argument against other defence priorities, particularly the Air Force's Air Combat Wing which was disbanded around the same time and the proposed purchase of F-16 Fighting Falcons cancelled. In 2005, however, the Auditor-General stated that it would have been possible for New Zealand to meet its commitments with fewer than the 105 NZLAVs it had purchased. The combat record of the NZLAV has been equally controversial. In 2009 eight NZLAV were sent to Afghanistan to support the operations of the NZSAS. In 2017 a video emerged showing how during the 2012 battle of Baghak in Afghanistan the inability of the NZLAV to elevate its main gun above 30 degrees had caused serious problems for the New Zealand troops.

The NZLAV has also been used in civilian operations. In May 2009 two were used to protect police officers in an armed siege when a gunman murdered a police officer in Napier. NZLAV were also deployed in the aftermath of the Christchurch earthquake in 2011. Nevertheless, New Zealand is currently trying to sell thirty vehicles from its NZLAV fleet. Originally twenty were up for sale, but an unnamed New Zealand army officer stated 'that the current fleet of twenty NZLAV was too small to generate a genuine interest'. At the time of writing no buyer has been identified.

OTHER USERS

In 2012 Colombia decided to purchase the LAV III and 32 serve in its mechanised infantry units. In 2016 it was announced that the Saudi Arabian National Guard has placed an order with GDLS for 900 LAV 6.0. Ex-US Army M1126 ICVs have also been purchased by the Royal Thai Army and sixty were delivered to Bangkok in 2019.

(below) Another ICV during Exercise Talisman Sabre 2017 held at the Shoalwater Bay Training Area in central Queensland, Australia. (Gordon Arthur)

STRYKER AT WAR: IRAQ

The Stryker made its combat debut in late November 2003 when 3rd Brigade, 2nd Infantry Division arrived in Iraq. The extensive training the brigade received at the National Training Centre in Fort Irwin, California, and the in-theatre upgrades (particularly in the form of slat armour to protect against RPGs) quickly proved their worth and the Stryker soon established itself as an effective defence against IEDs, as well as an capable weapon against insurgent forces. Much of the brigade's operations were of the non-lethal variety – working with the Iraqi authorities, communicating with the population and supervising contractors – and in this regard the Stryker brigade was the most visible and tangible Coalition presence in the region.

Commanders quickly learned that the current doctrine and organisation of the SBCT needed to be adapted to the tactical situation in Iraq. Initial problems were particularly acute in the Stryker reconnaissance troops, with their M1127 RVs. The lack of dismounted infantry in the reconnaissance units meant that one vehicle in each troop was frequently left behind at the Forward Operating Base to ensure each vehicle had sufficient security. By the end of their tour of duty the brigade's RVs were carrying a full squad of infantry each, utilising all ten bench seats in the rear of the vehicle. Other problems that quickly became apparent included the lack of an air-conditioning unit within the Stryker. One crewman observed: 'The Strykers are climate-controlled. That is to say, they are controlled by the climate. If it's 120 degrees outside, it's 130 to 160 degrees inside.' The introduction of cooling vents and personal cooling vests only partially solved this problem.

3rd Brigade's most notable action was Operation Black Typhoon in September 2004, designed to free the town of Tal Afar, some 60km west of Mosul in the north-west of the country, from insurgents. One incident in particular demonstrated the capabilities of the Stryker. On 4 September Iraqi insurgents succeeded in downing a Kiowa Warrior helicopter with an RPG. The Strykers succeeded in fighting their way through to the crash site and both recovered the damaged helicopter and rescued the injured pilots. Five other Americans and two Iraqis were injured and one Stryker damaged, while the insurgents were reckoned to have suffered over a hundred casualties during the action.

Over the coming months the Stryker's reputation as an effective weapon in counter-insurgency warfare grew. As 3rd Brigade had discovered, in Iraq the key to success lay, in part at least, in the number of dismounted infantry a unit could deploy. The SBCT could deploy many more infantry than the Heavy Brigade Combat Teams (HBCT) equipped with M1A1/A2 Abrams tanks and M2/M3 Bradley Fighting Vehicles. While tank commanders were grateful for the firepower of the 120mm gun of the M1A1/A2, collateral damage, such a factor in counter-insurgency operations, was much more easily minimised with the .50cal machine gun of the Stryker. During the troop 'Surge' of 2007-8 a second SBCT (4th Brigade, 2nd Infantry Division) was deployed to Diyala Province, joining the 3rd Brigade, 2nd Infantry Division which had arrived in the country in spring the previous year. In August 2007 the 3rd Brigade was

(below) An M1126 of 3-21 Infantry, 1st SBCT, 25th Infantry Division, during a patrol in Rawah, Iraq, in August 2005. Note the slat armour and improvised sandbag defences on the top of the hull. (US Army photo by Staff Sgt. Kyle Davis)

(above) Strykers from 4th SBCT, 2nd Infantry Division reach Contingency Operating Base Adder to refuel on their 'Last Patrol' to the Kuwaiti border in August 2010, marking the end of Operation Iraqi Freedom. (US Army by Pfc. Khori Johnson)

replaced by the 2nd Cavalry Regiment and in December a third SBCT, 2nd Brigade, 25th Infantry Division, arrived in country.

The deployments of 2007 saw the first employment of the M1128 MGS in Iraq. While this was not, as we have seen, without its teething problems, the arrival of the MGS only reinforced the growing reputation of the Stryker as an effective weapon against the insurgency. The Stryker was recognised by its crew as both durable and deadly. Sgt. 1st Class Scott Collum, an MGS commander in 1-38 Infantry, part of 3rd SBCT, 2nd Infantry Division, recalled how his vehicle was struck by an IED which blew out all the tyres and an antenna mount. Collum recalled how 'I was still able to drive the vehicle approximately 2,600 feet to a secure area. After replacing the tyres, a few caution messages were displayed on the computer. I powered down the MGS and powered it back up; all cautions were cleared and the vehicle was fully operational. I drove farther south and hit a second IED, and the same damage occurred. This time I identified the triggerman on the roof of a building 820 feet away. He ran out of a door on the top floor. With no tyres or communications and a few caution messages, we were still able to engage the spotter with twenty [7.62mm machine gun] rounds while on the move to eliminate the threat.' Collum was less impressed by having to leave the safety of his vehicle to fire the .50cal, but still considered the MGS 'the most lethal ground vehicle for an urban environment today.' Another advantage of the MGS was that the vehicle's thermal imaging sights could identify insurgents at ranges in excess of 2km, compared to only 100m for dismounted infantry. Perhaps the biggest advantage, however, that the introduction of the MGS provided was psychological. As one MGS crewman observed: '[The MGS] establishes fear in those who would try to do us or our fellow soldiers harm. The enemy is a lot less likely to attack if we are out there.'

In January 2009 the 56th Stryker Brigade Combat Team, the only National Guard Stryker unit, deployed to Camp Taji near Baghdad, to be joined by the 3rd SBCT, 2nd Infantry Division, stationed at FOB Warhorse in Diyala Province in August, and 4th SBCT, 2nd Infantry Division a month later (to western Baghdad). By the summer US forces were beginning to scale down their combat operations in the country, handing over security operations to Iraqi forces and police, yet the violence continued with numerous car bombings and other atrocities. The nature of the SBCT's duties had changed. In May 2010 4th SBCT helped support the Iraqi Security Forces in the presidential elections and began dismantling the infrastructure of the American war effort in Iraq. On 15 August 2010 the 4th SBCT began the symbolic 'Last Patrol' through from Baghdad to the Kuwaiti border. Three hundred Strykers, divided into four columns, took three days to reach Kuwait, marking the end of a war that had cost 4,400 American lives and some $478 billion.

Notes

1. The first deployment of the Canadian LAV III had been to Eritrea as part of the United Nations UNMEE mission in 2001 to monitor the ceasefire between Ethiopia and Eritrea. It also served with the NATO mission to Bosnia in the peacekeeping role.

(below) An M1132 equipped with a Lightweight Mine Roller and belonging to 1-24 Infantry, 1st SBCT, 25th Infantry Division clears 'Route Chicker' of IEDs during Operation Fairbanks in October 2011. Note the Duke anti-IED antenna mounted near the driver's hatch. (US Army photo)

STRYKER AT WAR: AFGHANISTAN

The Stryker/LAV III vehicle family had, of course, first seen combat in Afghanistan with the Canadians as part of the International Security Assistance Force (ISAF).[1] In August 2003 the 3rd Battalion, The Royal Canadian Regiment, deployed to Afghanistan as part of Task Force Kabul equipped with the LAV III. Between then and July 2011 the LAV III underwent various upgrades to meet the operational challenges and better protect the Canadian soldiers who served in it. Given its extensive use in Afghanistan, LAV III crew casualties were relatively light. Twenty-one separate incidents were reported involving LAV III which resulted in nineteen Canadian soldiers being killed and 67 wounded. The first came in November 2005 when a LAV III swerved to avoid on coming traffic near Kandahar and toppled over, killing one soldier and injuring four more. This incident gained a lot of negative press for the LAV III, especially as rollovers had been an issue during training in Canada, but in Afghanistan only six incidents (and five deaths) were due to road traffic accidents, while fourteen incidents involved IEDs and one an RPG attack. The fact that thirteen LAV III were lost to enemy action and a further 159 damaged, yet fewer than twenty soldiers lost their lives while operating them is testimony to the basic design of the LAV III and its ability to accommodate upgrades and additional protection to meet the current threat.

It was not until 2009, when President Obama ordered an additional 17,000 US troops to Afghanistan, that the US Army's Stryker Brigade Combat Teams were deployed to that country. The first was 5th SBCT, 2nd Infantry Division which arrived on 7 August. Based at FOB Frontenac, 30 miles north of Kandahar Airport, the brigade had only two weeks to prepare for its first operation: providing security for the forthcoming Afghan presidential elections. Faced with a resourceful and determined enemy, the brigade was assailed with IEDs and ambushes throughout its tour of duty. It had a hard baptism of fire: during its year-long deployment the brigade lost 37 killed and 238 wounded. In one incident, on 27 October, seven soldiers were killed in a single IED attack and by the end of 2009 21 vehicles had been lost to IEDs. An example of a typical incident was on 25 August when eight Strykers of 1-17 Infantry and a Mine-Resistant Ambush Protected Vehicle (MRAP) were sent from Frontenac to the Arghandab Valley to check on a reported cholera outbreak. En route the convoy was hit by an IED, which completely flipped one of the Strykers, landing it on its roof. The vehicle caught fire and four crewmen lost their lives, including the platoon commander and adjutant. The Taliban knew that US officers usually rode in the middle of columns and exploited this knowledge to deadly advantage. By contrast, US intelligence was at times poor. The commander of the 1-17 Infantry realised he was facing ten times the number of Taliban active in the Arghandab Valley than he had been led to believe. The rocky terrain and high hills of the Arghandab Valley also prevented the SBCT from bringing its full firepower to bear, leading them to be replaced in that area by 2-508 Infantry, 82nd Airborne Division, a light infantry unit better suited to exploit the terrain. Another issue was recovery of damaged vehicles. The tow bars fitted to the Strykers quickly proved inadequate and the Americans had to rely on Canadian armoured recovery vehicles. The recovery of

(above) The crew of an M1128 MGS of 4th SBCT, 2nd Infantry Division, watch as Afghan police pass out sweets to local children in Maiwand District in Kandahar Province in December 2012. (US Army photo by Sgt. Dayton Mitchell)

every damaged Stryker became a risky and dangerous tactical operation.

Fifth SBCT was relieved in June the following year by 2nd Cavalry Regiment. In Afghanistan the SBCT was usually dispersed with different battalions or squadrons serving in different parts of the country, much as they had done in Iraq. Third squadron, 2nd Cavalry Regiment, for instance, was deployed away from the majority of the brigade (based in FOB Lagman in Kandahar) to Combined Task Force Strike in Maiwand Province under the command of 2nd Brigade Combat Team, 101st Airborne Division. As such they took part in Operation Dragon Strike, designed to assert Coalition control over the so-called 'Heart of Darkness' area of Highway 1 in Kandahar. Just as in Iraq, it was the Stryker squadron's ability to transport large numbers of infantry into battle safely and quickly that proved vital during Operation Dragon Strike. By the end of December 2010 the US-led operation had succeeded in clearing Kandahar Province of Taliban forces.

In April 2011 1st SBCT, 25th Infantry Division, relieved 2nd Cavalry Regiment. Typical of the operational use of the Stryker in Afghanistan was the employment of 1-24 Infantry in Operation Fairbanks in October 2011. The objective was to clear and secure 'Route Chicken', the main road between Qalat and Mizan in Zabul Province of IEDs and build four checkpoints for the Afghan National Army. As well as physically locating and removing IEDs, regular patrols sent the message to the local populace that the route was, once again, open for business. Around the same time, 3rd SBCT, 2nd Infantry Division, deployed to Afghanistan but left its Strykers behind, operating MRAPs instead. The shortcomings of the Stryker in Afghanistan – the problems of deploying its weaponry in the terrain and its continued vulnerability to IEDs – were much publicised at the time, but as the brigade's deputy commander observed, "Even though we're a Stryker brigade with Stryker vehicles, the strength of the brigade is that we've got over 4,000 – it's an infantry brigade – armed soldiers inside of it."

In July 2013 2nd Cavalry Regiment began its second tour in support of Operation Enduring Freedom, the US mission in Afghanistan. Again deploying to Kandahar, it took responsibility for an area of the country previously covered by two brigade combat teams. One of their most frequent missions was to protect Kandahar Airport from indirect rocket attacks. 4-2 Cavalry, the brigade's RSTA squadron, became adept at searching out caches of rockets and their launchers, reducing indirect fire attacks on the airport by some 80 per cent during their tour of duty. By this time the US forces were also working in a much more integrated way with the Afghan National Army (ANA) and other local security agencies. During Operation Damavand I between 17 and 24 August Combined Task Force Dragoon helped the ANA clear enemy fighters from Panjwai, working with local police and village leaders to register voters for the upcoming elections in spring 2014. Once again the men and vehicles of the Stryker Brigade Combat Team had proved their worth. The training and professionalism of the men and women of 2nd Cavalry Regiment was just as important in winning the 'hearts and minds' of the local populace, transitioning to the Afghan military and civilian organisations and preparing the ground for elections, as it was in engaging the Taliban in combat.

FURTHER READING

There is surprisingly little in print about the Stryker family of vehicles. Gordon Rottman's *Stryker Combat Vehicles* (Osprey New Vanguard, 2006) is a handy introduction, but now rather dated. R.P. Hunnicutt's *Armored Car: A History of American Wheeled Combat Vehicles* (2015) provides an exhaustive history of wheeled AFVs in American service up to and including the Stryker family. The most informative publications, at least from a modeller's point of view, are those by Ralph Zwilling and Carl Schulze. Their *Stryker: Interim Armored Vehicle*, was published in 2007 in Concord Publications' MiniColor series. This is a great collection of images of early Strykers, particularly those of 2nd Cavalry Regiment in Germany and in Iraq. Ralph Zwilling's volumes for Wings and Wheels Productions are also invaluable companions for anyone interested in the Stryker. *Stryker in Detail: US Stryker Interim Armored Vehicle Family Part One* (2007) and *Stryker Family Upgrades: 2008-2014 Modernizations (2015)* are both superb books, the former looking at vehicles in Germany and Iraq, while the latter concentrates on the vehicles of 2nd Cavalry Regiment. *Stryker MGS in Detail: M1128 Mobile Gun System* (2010) is a photo manual, again concentrating on 2nd Cavalry Regiment's vehicles. Ralph has also several shorter articles on various Stryker variants in Tankograd Publishing's quarterly *Militärfahrzeug* magazine. These include the February 2019 issue on the M1296 ICVD and April 2018 on the MEHEL Stryker prototypes.

The M1128 MGS is also the subject of an excellent photo book by Sabot Publications, *Stryker MGS: the Mobile Gun System in Combat* (War Machines 02. 2017). Sabot Publications' have also published a similar volume on the *Stryker Dragoon, M1296 ICV Dragoon* (War Machines 06, 2019). Publications on the Canadian LAV are similarly few and far between: *Canadian LAV III in Afghanistan* (Modeller Photo Assistant 1, 2008), written by Anthony Sewards and Miroslav Hraban is published by Real Model and is a great collection of images, invaluable for anyone tackling a Canadian LAV project in miniature.

Thanks to Mark Smith, Carlos Blanco, David Chou, Sean Lynch, David Parker, Ramón Segarra Guerrero, Patrick Winnepenninckx, Slawomir Zajackowski, Gordon Arthur and Ralph Zwilling, for contributing ideas, photographs and models to this volume. This volume would not have been possible without the support of the Defense Visual Information Distribution Service (DVIDS).

(right) An M1130 CV of 1st SBCT, 1st Armored Division, fitted with the Warfighter Information Network - Tactical (WIN-T) equipment, tests the network alongside British soldiers from the Scots Guards in Fort Bliss, Texas, in September 2015. (US Army photo by David Vergun).

An M1126 ICV of 1-24 Infantry, 1st SBCT, 25th Infantry Division crosses Jarvis Creek in the Donnelly Training Area, Alaska, in October 2018 during Exercise Artic Anvil, a multi-national training event held with Canadian forces. (US Army photo by John Pennell)